CW01658559

# THE ROYAL NAVY IN HONG KONG

## SINCE 1841

Kathleen Harland M.A.

# Contents


| | | Page |
|---|---|---|
| Chapter 1 | Establishing of the Naval Base 1841-1856 | 3-8 |
| Chapter 2 | The Development of the Dockyard 1856-1902 | 9-18 |
| Chapter 3 | The Year 1878 | 19-26 |
| Chapter 4 | Expansion through Reclamation | 27-34 |
| Chapter 5 | The Dockyard under Attack—December 1941 | 35-42 |
| Chapter 6 | The Return—August 1945 | 43-53 |
| Chapter 7 | The Closure of the Dockyard—1957-59 | 54-57 |
| Chapter 8 | The Fifth HMS Tamar | 58-69 |
| Chapter 9 | The Hong Kong Chinese Locally Entered Personnel | 70-73 |
| Chapter 10 | Health | 74-79 |
| Chapter 11 | The Navy in the New Territories | 80-83 |
| Chapter 12 | General Contribution to Colonial Life | 84-88 |
| Chapter 13 | Pirates on the Chinese Main | 89-102 |
| Chapter 14 | Recent Events | 103-112 |

CHAPTER 1

## THE ESTABLISHING OF THE NAVAL BASE 1841 - 1856

The Royal Navy came to Hong Kong in defence of the opium trade. It would not, of course, have been stated quite as baldly as that at the time; rather would it have been said that the Navy was merely protecting Britain's commercial interests. That the main constituent of that commerce was opium was neither here nor there. Since 1757, when Canton was made the centre of trade with the west, enterprising British merchants had found a ready sale for their product among the Chinese population. Although the poppy was grown in China, most of the opium came from British-ruled India, which supplied an ever-increasing demand in the 18th and early 19th centuries in spite of the fact that the whole business was illegal. Prohibited, as it was, by the edict of successive Chinese emperors, who took the very reasonable view that smoking opium was injurious to their subjects' health, and should be banned. For many years, however, these well-meaning laws were ignored, until the fateful day of March 10th, 1839, when one, Lin Tse-hsu, arrived in Canton as Special Imperial Commissioner for the suppression of the opium trade. He acted quickly: within a week, having taken hostages, he forced the surrender of all British-owned opium in the port (20,291 chests of it!) whereupon the merchants, in high dudgeon, departed first for Macau and later, when refuge was refused there, for the safety of vessels anchored in the bays of neighbouring islands of which Hong Kong was one.

It was not long before the home government was drawn into the dispute. On September 7th the merchants wrote asking for prompt intervention and retaliatory action against the officials in Canton. With the strong commercial interests represented in the Parliament of the day, their plea was certain to be heard. The distance involved, moreover, ensured that "intervention" would have to come in the form of the Navy's troopships

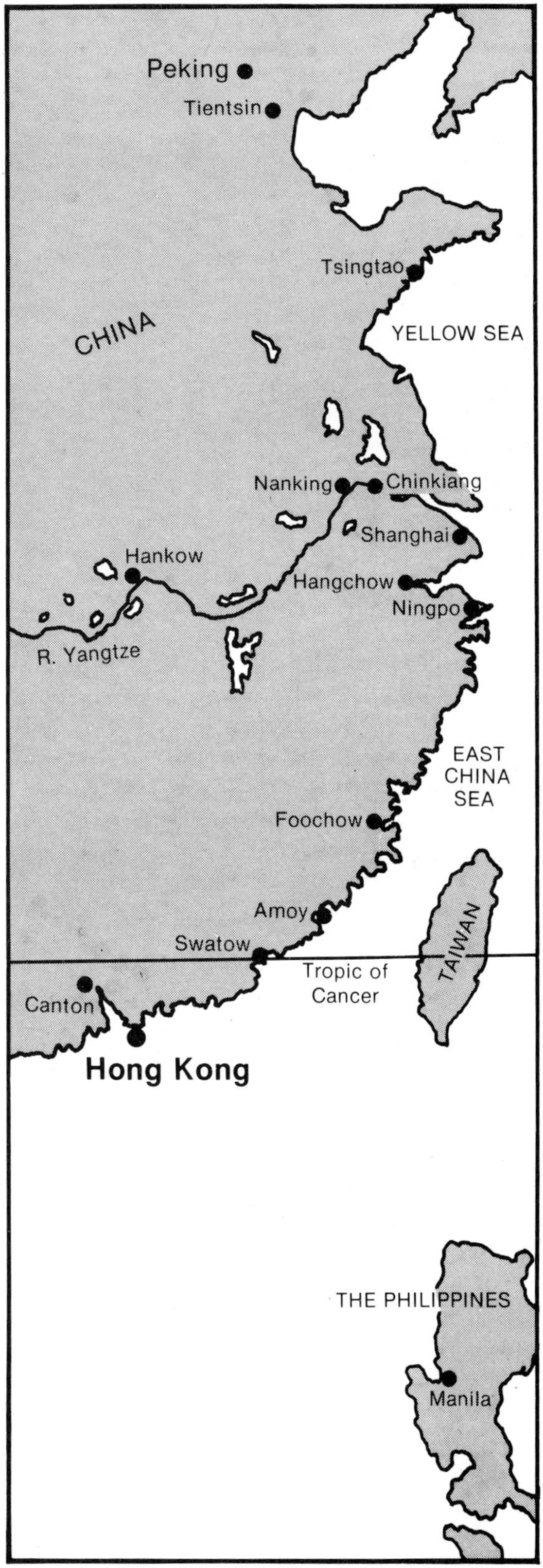

**The Conference at Chusan, 1840. On the basis of an agreement signed here, the Navy made the first landing on Hong Kong Island.**

and men-of-war, while the geographical position of Canton on the Pearl River meant that naval bombardments would have to form part of any future "retaliatory action". By June 1840 there were seventeen warships in Hong Kong harbour, among them three line-of-battle ships, including H.M.S. Melville (later relegated to a hospital hulk) and twenty-seven troopships. The expedition was commanded first by Rear-Admiral George Elliot and later by Commodore Gordon Bremer, while the Admiral's cousin, Captain Charles Elliot was given the grand title of "Plenipotentiary", which simply meant that he was in charge of the negotiations with the Chinese, a post to which he was suited for, as Superintendent of Trade in Canton since 1834, he was aware of the intracacies of the situation.

The anticipated naval actions took place in January and March the forts controlling the Bogue (the narrow channel leading to Canton) were bombarded—and then occupied by troops. This was enough to bring the Chinese to the conference table and a convention was agreed at Chuenpi (though never signed) on the basis of which Elliot allowed a naval party under Bremer to land and take possession formally of Hong Kong island on January 26th 1841.

Their actions, however, did not meet with the approval of the home government, and Elliot and Bremer were recalled in disgrace. Admittedly they were luckier than their Chinese counterpart who was sent to Peking in chains. Elliot, too, gave his name to what was once called "the most romantic valley on the island", Elliot's Vale, or, as it is now known, Glenealy.

A new team, headed by Admiral Sir William Parker and Sir Henry Pottinger arrived, determined to find a solution to the problem which was, in essence, to establish a suitable base from which the English could carry on trade unhampered by Chinese interference. This necessitated the use of more force: between October and December 1841 a combined naval and army expedition took Amoy, Tinghai, Chusan and Ningpo before winter brought an end to the campaigning season. When spring came a Chinese counter-attack on Ningpo was repulsed, the fleet bombarded the Woosung Forts at the mouth of the Yangtze River prior to troops occupying them, which left the way open to Shanghai. The fleet then continued up river and after capturing Chinkuang and blocking the Grand Canal, lay off outside Nanking. Although in comparison with naval exploits of the Napoleonic era these actions were on a small scale, they achieved the desired object: to make the Chinese negotiate. By the Treaty of Nanking of August 1842, British merchants gained the freedom of trade they had long wanted, and another jewel was added to the imperial crown: the island of Hong Kong. To quote from the treaty itself:

> "It being obviously necessary and desirable that British subjects should have some Port whereat they may careen and refit their ships when required and keep stores for that purpose, His Majesty the Emperor of China cedes to Her Majesty the Queen of Great Britain . . . the island of Hong Kong, to be possessed in perpetuity by Her Britannic Majesty, Her Heirs and Successors."

**ARRIVAL**

The first naval officer to set foot on Hong Kong island was probably Sir Edward Belcher, Captain of H.M.S. Sulphur. He recalls the landing in his book, "Narrative of a voyage round the World, performed in H.M.S. Sulphur.":

"On the return of the Commodore (to Macau) on the 24th (of January, 1841) we were directed to proceed to Hong Kong and commence its survey. We landed on Monday the 25th 1841 at Fifteen minutes past 8 a.m. and being the bona fide first possessors, Her Majesty's health was drank (sic) with three cheers on Possession Mount. On the 26th the squadron arrived; the marines were landed, the union (sic) hoisted on our post, and formal possession taken of the island by Commodore Sir J.J.G. Bremer, accompanied by the other officers of the squadron, under a feu-de-joie from the marines and a royal salute from the ships of war."

His somewhat laconic account makes it all sound very much in a day's work!

The place where he landed came to be known as Possession Point, and even today there is a Possession Street which runs from Queen's Road Central up to Hollywood Road. Bearing in mind that the sea formerly came right up to Queen's Road, it is probable that the lower or northern end marks the landing site, but strangely, no memorial has ever been erected to commemorate the occasion. The captain himself is well remembered: in Belcher Street, Belcher's Gardens (an apartment block) and Belcher Bay, along the shore of Kennedy Town. His ship, from which the first survey of the island was made, gave its name to Sulphur Channel between Green Island and the western tip of Hong Kong island.

It would appear that the second landing was made further east than the first because, after the ceremony, Captain Belcher took his first readings near Morrison Hall and named the principal surrounding high places. Once again the event is remembered in a street name, "Observation Place", which led from Hennessey Road to Morrison Hill Road. In 1911 the street was re-named Tin Lok Lane which means Happy Heaven Lane, thus keeping up the old idea of the stars and planets from which Belcher's calculations were made, and referring also to Happy Valley, situated at one end of the street.

**SETTLING IN**

Very soon after these ceremonies of possession, in April 1841, the naval authorities, along with their military and civil counterparts, began to erect buildings along the foreshore. The first naval store sheds were at West Point and seem to have taken the form of matsheds, a rough type of wooden warehouse with a mat roof. These were erected a few miles west of Possession Point. They may well have been badly hit by the typhoon of the 21st July, which certainly caused destruction among shipping in the harbour. Very early in the colony's history the line of Queen's Road, which was to become one of its most famous roads, was clearly marked out. This is probably because it followed the line of an old track used by the crews of junks who, when the wind failed, had to pull their craft along the coast.

A study of early maps reveals that between 1845 and 1855 major construction was carried out at a site slightly westward, and also that at the existing site two new buil-

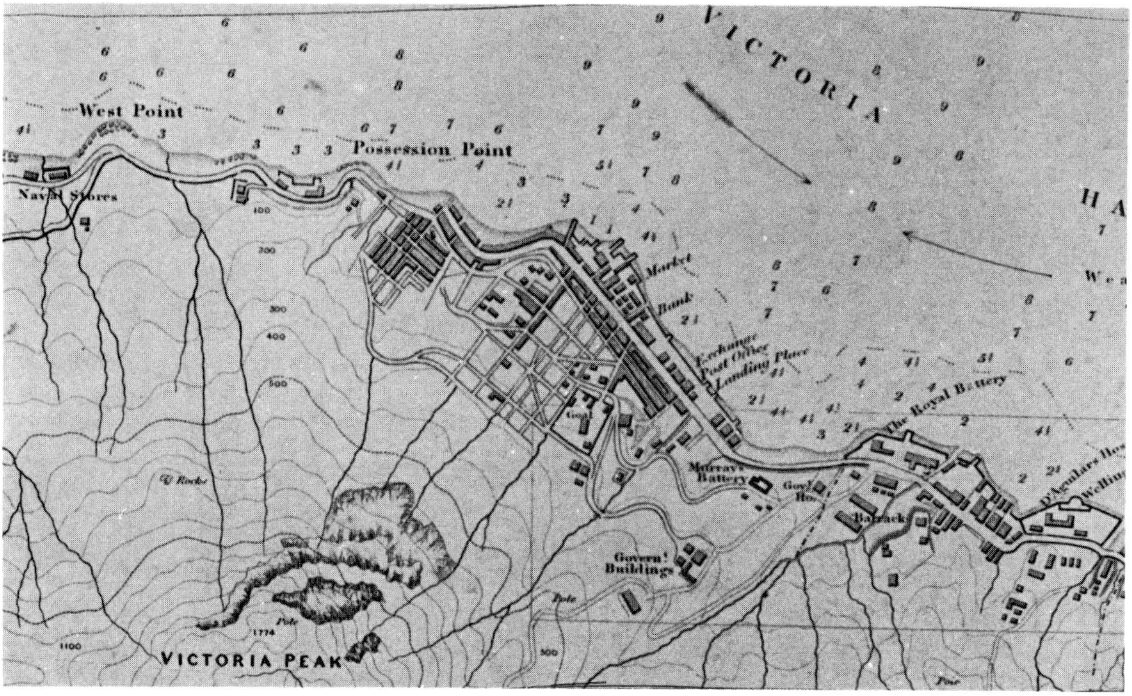


**Early Days — Four Years after landing. Note the Navy Stores sheds at West Point. The outline of the famous Queens Road can already be seen.**

dings were added. The first Naval Storekeeper and Agent Victualler was appointed on the 21st March 1842. This was a certain Thomas McKnight who was to hold the post until the 12th October 1849, when he was relieved by Commander Charles Parker as Naval Officer in charge of Stores. He, in turn, remained until April 1858, when reorganisation took place. The fact that two such posts were combined in one person in the early days points to the fact that naval dispositions were still on a small scale. Mr McKnight may well have commuted between the original West Point site and a new one in Central, taken over by Admiral Parker late in 1842 or early in 1843. It is not clear why this additional site, which was, in fact, to prove the permanent one, was occupied, but it could well be that Admiral Parker was spurred on to his action by the example of his military counterparts: a Major Aldrich, sent out by the War Office in late 1842, produced an ambitious scheme for a military cantonment to both the east and west of what is now Garden Road, and it may well be that the good admiral did not wish the Senior Service to lag behind in acquiring a central position in the infant settlement of Victoria. Certainly, his action annoyed the governor, Sir Henry Pottinger, who had hoped to save this valuable area for commercial purposes, but his request to the Navy to move was refused; similar requests by future governors were to meet with an identical response.

While store depots were thus quickly established on shore, the Navy also made use of ships moored in the harbour: H.M.S. Minden, originally a hospital ship in the very early days, was used as a store ship from 1846, and during the Second China War (1857-59) the Princess Charlotte was sent out to act as a Receiving Ship. This pattern was continued for in 1875 H.M.S. Victor Emmanuel relieved the Princess Charlotte, only to be replaced in turn by H.M.S. Tamar in 1897, the ship which has given her name to the present naval base, and which was scuttled in December 1941 prior to the Japanese occupation.

A Receiving Ship seems to have had certain functions, one of which was to provide office accomodation. Certainly, the Commodore used it as his administrative base in the nineteenth century. Salutes, on occasions like the Queen's Birthday, were fired from the ship, not from the yard, court martials were held on board at irregular intervals, and when Governor Hennessy visited Commodore Watson "and stayed to tiffin", as he did on the 23rd April 1878, it was to HMS Victor Emmanuel that he went and not to any establishment ashore. For certain months of the year, moreover, the Commodore and his family lived on board. Plays and the musical entertainments so beloved of the Victorians were performed on the ship, so there must have been a sizeable wardroom or hall with catering facilities, as many civic dignatories used to be invited along. The Receiving Ships too must have been "home" to many officers and men. This is certainly true of H.M.S. Tamar in the present century, when she was moored alongside the west wall of the basin, and may well have applied earlier because, in 19th century maps of the naval yard, there is only one reference to a married quarters block on shore. The inference is that most men, on their tours of duty abroad (and it is only relatively recently that the "married accompanied" draft has come into being) would live and work on board ship.

**An idea of the emptiness of the harbour in the 19th century can be gained from this photograph. H.M.S. Victor Emmanuel is in the centre (without masts).**

H.M.S. Tamar as she looked when in service as a troop ship to Africa and the Far East.

CHAPTER 2

## THE DEVELOPMENT OF THE DOCKYARD 1856 - 1902

The earliest reference to the naval yard on the island is probably that of July 22nd 1856. It is contained in a memorandum concerned with a further transfer of land (there had been one in 1854) by the military to the Admiralty. The land involved would appear to be that section on which the Mass Transit Admiralty Station is now built. It should be remembered that for most of its history, until the major handover of land to the civil government in 1959, the southern boundary of the dockyard was Queen's Road; what did vary was its extent to the east and west. On both sides the yard was bounded for many years by Army property: Wellington Barracks to the east and the North Barracks to the west. In these transactions of 1854 and 1856, therefore, the yard is expanding at the expense of the North Barracks, used by the artillery. The reason must have been that extra space was required due to the surrender, also in 1854, of the ground at West Point. A map dated 1855 indicates that the store sheds there, are in the process of demolition. Presumably, as the infant city of Victoria began to grow, the advantages of the central site became obvious to the naval authorities. No doubt new buildings were put up to accommodate the stores, or possibly old ones were utlised.

The expansion of the premises entailed an increase in personnel. For the first time, in the Navy List for 1859, there is an entry for Hong Kong under the heading "H.M. Victualling Yards" as well as in the section

**An unusually close view of the Dockyard buildings in the 1880's, showing the two piers in use with the sheer legs used for bringing stores ashore—and the launches from the gunboats in harbour.**

A closer look at the dockyard buildings in the later 19th century. The main entrance was between the two large buildings in the foreground, which fronted Queens Road, of which a small part can be seen. To the left are the swimming huts at the end of a pier belonging to the Army's North Barracks.

listed "Dockyards". The victualling yard must have been established by 1858, probably on the former army land, as this was when its first "Paymaster in Charge of Stores" was appointed. From then on, the dockyard and the victualling yard are listed as two separate establishments, each with its own officials. The dockyard officials would be responsible for all the machinery and spare parts necessary to maintain and repair ships, while the victualling yard storekeepers would hold dry goods and foodstuffs. This reorganisation of 1859 would have been a consequence of the renewed hostilities with China during 1856-58 (known as the Arrow War). War would have meant increased demand for supplies, ammunition and spare parts. Although the warships were operating in the Pearl River Delta they used the Hong Kong yard as their base.

The first reference to an anchorage being set apart for men-of-war is contained in a memorandum of January 1860. No details are given, but presumably it was opposite the yard, and out in the main channel. Once anchored, however, it can have been no easy matter to get ashore. Pedder's Wharf was the main landing stage for the whole waterfront. A passage from the Hong Kong Telegraph of September 3rd 1883 describes it thus:

"It is a miserably attenuated wooden structure. In size it is not one half large enough for the traffic of the city. From morning to night the flight of wooden steps are crowded with steam launches, men-of-war gigs and cutters . . . and native craft of every description."

Sailors, then eager for some relaxation ashore, must have had to manoeuvre as skilfully in the shallow waters as they did in deeper, and once on land, have had to elbow their way through the press.

The landing stage in question was named after Lieutenant William Pedder, R.N. of H.M.S. Nemesis, who in August 1841 became the colony's first Habour Master, Magistrate and Surveyor of Shipping. His office was on the slope to the east of Wyndham Street and his wharf was below, at the end of Pedder Street, approximately where the traffic lights now stand in the middle of Des Voeux Road. This was the original waterfront but the site has long been left far inland by reclamation. In the old days, Pedder's Wharf was a definite locality of the city, and several business firms had their address as such, but when the great reclamation scheme in Central was completed at the turn of the century and a new pier constructed, the latter was named Blake Pier after the then governor. When the governor left the colony in 1908 there were complaints in the Press about the loss of the old name, but by then it was too late to do anything about it. Pedder Street, however, still retains the link with one who, among many naval officers, made a distinctive contribution to the colony.

**KOWLOON**

The acquisition of the Kowloon Peninsula by Britain in January 1861 meant additional premises for the Navy. It is thought that an English naval officer, Captain W.K. Hall of H.M.S. Calcutta was the first to appreciate the importance of the peninsula. In a letter of March 2nd 1858 he pointed out that the peninsula would give much needed sea-frontage, that Stonecutter's Island could be useful as a quarantine area and for strengthening the defences of the island, and suggested finally that the naval yard should move to a site on the mainland. The Governor, Sir John Bowring, liked his proposals and put them forward in his report to the Colonial Office. In the event, after Kowloon was ceded, a Board was set up to determine how the land was to be apportioned, on which Captain Borlase represented the Admiralty. The lion's share went to the Army but a site was acquired by the Navy and used for coal storage.

Several times during the late 19th and early 20th centuries it was suggested that the Navy should move all its facilities to the Kowloon Yard. Successive governors, with their eye on the prime central location on the island, would certainly have liked this, and the plan was also favoured by some senior naval officers, but eventually it foundered on

Whitehead Torpedo Workshop, Kowloon, January 1895.

1st class torpedo boats in the Camber at Kowloon, 1888.

**The Lamont and Hope dry docks at Aberdeen in 1868, when they were the only docks available for use at Hong Kong.**

the rock of Admiralty opposition. The site remained naval property until 1959 when, by the agreement of the 4th October, the entire yard was handed back to the civil authorities.

With two yards now in existence it was obvious that administrative posts would have to be created commensurate with increased responsibilities. In 1861 a Master Attendant of the Yard and Accountant were appointed while the commander of the Receiving Ship, at that time H.M.S. Princess Charlotte, was made "Naval Officer in Charge of the naval establishments". Five years later the post of "Commodore Hong Kong" was created, one which was to last for over a century. In 1880 an Inspector of Machinery was appointed for the first time, an old-fashioned sounding title which was changed to the more modern Chief Engineer in 1890. In 1884 he had under him a foreman, three engineers (for re-venting heavy guns), boatswain and carpenter who were joined the following year by a Constructor (for re-fits) and an Inspector of Shipwrights.

The naval establishment on the island had always had a military guard, probably because it was sandwiched between the North and Wellington Barracks, but with expansion and increased funds available, this was deemed unsuitable. It was withdrawn in February 1866 when Dockyard Police were engaged. In the next year the Commodore asked the Surveyor General to have the boundaries of the naval property measured. This may well have been through fear of possible encroachments from the Chinese area of shops and houses to the east of the yard, and may indicate the absence of proper boundary

**Aberdeen docks destroyed by a typhoon in 1874. These docks were frequently used by naval vessels.**

adjoining areas was evident. This is shown by correspondence of 1877 over Army claims that the Navy was encroaching on its property. As has been already stated, part of the North Barracks had been acquired by the Navy in 1854 and '56. In the year in question the naval authorities had built a cook-house on what they considered to be their own boundary line, only to find the Army claiming it as its own. Claims and counter-claims were settled by the payment, for a period, of the sum of one dollar, by the Navy to the military as a form of rent or "annual acknowledgement". The problem of space, as will be seen, could only be resolved either through a move or by seaward expansion. These two points were to exercise both the naval and civilian authorities from the 1880's until their resolution at the turn of the century.

walls. Space, indeed, always seems to have been at a premium, and pressure from

## POSSIBLE MOVES

A quadratic correspondence on this subject, involving the local naval officers, the colonial government, the Colonial Office and the Admiralty was begun in December 1875. The particularly severe typhoon of the previous year had caused extensive damage along the praya (water-front road). In his plans for reconstruction, the Governor, Sir Arthur Kennedy, suggested that the praya be extended eastwards, in front of the naval and military areas. These areas, sloping down to the water as they did, almost divided the colony in two: Queen's Road was the only link between the Central and Wanchai districts.

Obviously, the civil government would have liked another link road and indeed, the Navy would not have objected as the harbour was silting up in front of the yard leaving their landing stage almost high and dry at low tide, which sometimes entailed dragging stores over a stretch of mud. The proposed sea wall would be built far enough out to be in deep water and special entrances in it would be devised for boats and even large ships. As often, however, the plan foundered on who was going to supply the money. Kennedy had hoped that the Admiralty would bear the cost; the Admiralty felt that, since the scheme would mainly benefit the civil sector, the colony should provide the funds. No more was heard on the subject though the building of Kennedy Road, on the southern boundary of Victoria Barracks, did go through by way of a consolation prize, to provide a second major link with Wanchai. In 1885 the possibility of building a praya was ended by the Admiralty's decision to extend the yard seawards through reclamation.

The first suggestion that the Navy should actually vacate its site was made in September 1881 when a Colonel Crossman, R.E. wrote to the Governor proposing the transfer of the military and naval establishments to Causeway Bay, and enclosing a plan, which still exists, which he had drawn up to this effect. As before, the Hong Kong government would have been only too willing but in March 1882 the Admiralty notified the Commander-in-Chief, Admiral Willes, that it could not agree. In March 1887 a shift to Kowloon is mentioned for the first time at the initiative of the civil government. The Governor offered Admiral Hamilton land of equal value in Kowloon but although Commodore Morant was quite agreeable to the transfer, the Admiralty in June pointed out the realities of the situation when it said it was not prepared to purchase the land. Sir William des Voeux in January 1888 pleaded that to bear the transfer alone would involve the colony in "ruinous expense", and so yet another stalemate was reached.

In the following years many letters passed back and forth with the same ideas being bartered around but no changes were made until 1896 when the yard's southern boundary was re-drawn to allow for the construction of what is now the east-bound carriageway of the Queen's Road to Wanchai.

In the same year the naval holding expanded to the east. There, like a wedge between it and Wellington Barracks was a parcel of land on which two rows of Chinese dwellings and shops had been built from the early days. The site was very close to the Ordnance Stores, too close for Admiral Salmon's liking, for he had pointed out the great danger involved should the houses catch fire. They were owned by a Mr Edmund Sharp, who offered to sell out for £70,000 or to lease the land at an annual rent of £3,000. The first offer was accepted in April 1896.

On the opposite side of the naval base was the North Barracks. The naval authorities had had their sights trained on this target from the 1870's because they wanted it as accomodation for ship's crews in lieu of a Receiving Ship. Not surprisingly, considering its age, H.M.S. Victor Emmanuel was proving unhealthy, and so the Navy had to go house-hunting. The North Barracks seemed a most desirable residence; it was simply a matter of removing the owner-occupiers. In 1887 this was finally achieved: the Army acquired the Mount Austin Hotel, though in deference to the wishes of the many residents it postponed its occupation for several months. This year, therefore, saw the conclusion of the yard's western expansion up to Murray Road. Where the multi-storey car park (and beneath it, the Transport Office) are now, was the western corner of the naval base, and it remained until the major handover of land to the government in 1959.

The final chapter in the "movements question" was written in May 1901, when the then commodore, Commodore Powell, forwarded home a proposal by Sir Paul Chater to move the base to the west side of the Kowloon Peninsula. Sir Paul, a wealthy merchant and member of the Legislative Council, was the driving-force behind the great reclamation scheme in Central, which had been begun in 1890, extending the land from what is now des Voeux Road (then the water-front road) to the present shore-line. It would have been neat, and possibly aesthetically satisfying, if this could have been continued right along to Causeway Bay. In weighing up the pros and cons of his suggestion, the Commodore admitted that there was a need for further expansion "and if the present site were retained it would eventually be necessary to acquire more property", but on the other hand, dredging would be necessary at the Kowloon site. If there had to be a move, Rocky Peak at Lyemun or Hung Hom Bay at Kowloon might be more suitable. One cannot help thinking that successive naval officers threw up these suggestions partly to fog the issue, very much in the way that one of their own ships might throw up a smoke-screen. At any rate, both these plans were rejected by the Admiralty in October. Presumably the thinking was that it was pointless to move from a site which was basically very suitable simply to improve the view or satisfy the commercial interests of Hong Kong businessmen.

**An aerial view of Aberdeen dockyard and village. This later view is in striking contrast to the desolate scene of the earlier photograph.**

A 19th century view of Victoria. The old cricket ground, with its pavilion on the water-front (where Chater Road now is) can be clearly seen. To its right are the North Barracks, the Naval Base and Wellington Barracks. The large white ship is H.M.S. Tamar.

CHAPTER 3

## THE YEAR 1878

Inserted into each of the four main gates at the entrance to the modern H.M.S. Tamar is an oval medallion containing an anchor with its cable attached, and underneath, the number 1878. It is not known when this device was designed and its origins must remain uncertain. The date, however, probably refers to the year in which, according to government records, H.M.S. Tamar first visited Hong Kong.

1878, then, a year from the high Victorian era, would be a suitable time to stop and view the Navy's activities, to see what pursuits it engaged in at that time, what matters held its attention.

The naval year always began with the Victorian Rowing Regatta. Rowing was a favourite pastime, certainly more so then than now, perhaps because the harbour was much quieter and leant itself to such sport. The size of boat used in the races varied somewhat: smaller ships like the gunboats Moorhen and Swinger raced their gigs, while larger vessels such as Victor Emmanuel used their whalers; at the beginning of the year Tamar scored several successes with her cutters and had enough to put in several teams. The course was laid out between wellknown landmarks: Victor Emmanuel was usually the starting point, then off, down and round one of the army hospital ships, like H.M.S. Meanee, and back. For variation, there would sometimes be a straight dash from the receiving ship to the coaling sheds at Kowloon.

The ships which were in harbour at the beginning of the year, apart from Tamar which soon left for Singapore, were the receiving ship, H.M.S. Victor Emmanuel, flying the flag of Commodore Watson, its paddle dispatch boat, H.M.S. Vigilant, three gunboats, the Lapwing, Curlew and Growler fitted with three guns apiece and engines of between 120 and 160 H.P., and finally H.M.S. Audacious, the flagship of the China Squadron, paying its Christmas visit to the colony. It was

probably to mark this event that the seamen and marines from all the ships had landed at the Yard promptly on January 25th at 9.15 a.m. "and proceeded, accompanied with the field pieces of H.M.S. Audacious and the band, to Happy Valley, taking the road by the Praya East." It is interesting to note that there cannot have been space in the Yard itself for such an event. At all events the Admiral was reported pleased with the various manoeuvres carried out.

Four days later Tamar returned from Singapore with the 74th Highlanders, who were relieving the 28th Regiment. Her officers were just in time to attend the Saturday night entertainment organised on board Victor Emmanuel, possibly in celebration of the Chinese New Year which began that day, although the entertainment itself could not have been more English, for according to the newspaper correspondent, it began with "songs and dances by some of the Sailors. Lieutenant Needham gave a reading, "The Vulgar Little Boy" with great spirit, and kept the audience in roars of laughter. The entertainment was concluded with the well-known farce "Box and Cox". The characters were very well sustained by Lieutenants Fenwick and Barnard who were very well up in their parts. The performance was given on the quarter deck, upon a platform made for that purpose, with scenery painted for the occasion. A great number of military officers, civilians and about six hundred sailors were present, who greatly appreciated the performance." So ends the newspaper account; it is left to the imagination to supply a picture of the ship, moored off-shore, lighted up in welcome against a harbour much darker than it is now, of the splash of oars as the guests were brought out, of boat cloaks stirring in the breeze, of footsteps on the boarding ladders, of English laughter and English humour drifting across the oriental scene.

The next afternoon, having recovered from the night's entertainment, or possibly refreshed by it, the Navy sallied forth to play the Hong Kong club at cricket. It was a Wednesday and probably then, as now, constituted a half-day for sporting activities. Cricket, of course, was a popular sport, played in the colony from the very first during the cooler months. Overall the Fleet seems to have held its own against its three main "enemies", the Cricket Club, the Hong Kong Club and the Army.

Nor were the festivities yet finished. While the local Chinese were still celebrating the New Year, the Naval Ball was held on February 5th at the City Hall in the presence of Governor and Mrs Hennessy and with music provided by the bands of Audacious and the 74th Highlanders. It was obviously an occasion to remember, one which certainly dazzled the reporter from the "Daily Press" who surpassed himself in his choice of language brilliant enough to describe such a brilliant event. "The City Hall assumed a very gay appearance for the occasion, testifying to the hearty welcome of the gallant hosts. A field piece stood on each side of the hall; men-of-war's men lined the staircase and formed a guard of honour on the arrival of the ladies; a Gatling gun occupied the centre of the first landing and was supported on each side by a boat gun and piles of polished projectiles. Two inflated diving dresses, more curious than ornamental, stood against the wall. The ballroom was gorgeous with decoration, every gas bracket being surrounded by a blazing sun of bayonets and ramrods.

The supper room was decorated with equal taste. The verandahs were enclosed with sails and mats and were draped with flags. A very pleasing and novel feature was the formation of an improvised conservatory at each end of the front verandah, reflecting much credit upon the designer, Lieutenant Shaw, R.N. A fountain embedded in rockwork played in the centre of each, while ferns and plants tastefully arranged and lighted by ships' lanterns rendered the whole the delightful realisation of a happy thought. The dance programmes were especially designed for the occasion and appropriately bore the Admiral's flag and the Commodore's burgee and the figures of war vessels."

After reading this account one can only think of the tremendous amount of preparation needed to produce such a result: polishing a pile of projectiles is not finished in an hour or even a morning! Of course, those were "spacious" days when there was no shortage of man-power, and although the practice of naval hospitality is still very much alive, in today's smaller, more stream-lined force, there is less time and perhaps even inclination for such a lavish display.

The Commander-in-Chief of the China Station, Vice Admiral Hillyar, had now been in Hong Kong since before the New Year, carrying out a series of inspections. On February 19th he took out those of his squadron which were in harbour—his own flagship and four gunboats—for tactical manoeuvering and target practice. A reporter was presumably borne on one of the ships, for the "Daily Press" had this to say: "The Fleet exercised at general quarters during the day, firing at a rock distant about 2,800 yards and made excellent practice, all the shot falling within three feet of each other. At 11 p.m. the squadron exercised at night quarters, the guns being cast loose and three rounds were fired in three minutes. The 20th being very misty, the fleet remained at anchor in Tytam Bay. Yesterday morning at nine they weighed anchor and proceeding out of the bay they manoeuvered under steam and sail, the evolu-

**The City Hall, a typically gracious, colonial-style building, seen from the cricket ground, (about 1880). Immediately behind the pavilion to the right was the water-front.**

tions giving great satisfaction to the Admiral. At noon the squadron proceeded for the anchorage, with the exception of the Magpie (the only one of the gunboats completely dependent on sail) which went to Aberdeen to be docked."

A few days later some light relief was supplied by a game of baseball, or rather, as the "Press" had it "a game at Base Ball", between the Cricket Club and officers from the English ships and the U.S.S. Tennessee. From the newspaper account it would seem that this was the first baseball game to be played in the colony. It says, quite rightly, that "Base Ball is not unlike Rounders. Its features give plenty of scope for the dexterity required on the cricket ground, and in batting and fielding the cricket players soon showed themselves quite at home." The shore team, in fact won, which must have been a nasty shock for the American team, diluted though it was by the presence of the Royal Navy. The allies did no better later in the year, in October, when the cricketers again routed them.

In March Vice Admiral Hillyar stepped up the training programme for his crews, in view of the political situation at home. In eastern Europe Russia was at war with Turkey in an attempt to gain territorial concessions, Britain and France were trying to stop her, initially through diplomatic channels, but, as these seemed likely to fail in the early part of the year, there was a possibility, though a somewhat remote one, that the China Squadron might have to repel a Russian attack from the north. On the 1st April it was announced that "in consequence of the warlike telegrams recently received we understand that all the munitions of war have been carefully surveyed." Three days later Audacious, with Magpie in tow, left for a cruise "under sealed orders." The newspaper states that their final destination is the anchorage outside Vladivostok harbour, but subsequent reports place them never further north than Yokohama. Those left behind battened down the hatches by moving the Victor Emmanuel from her mid-channel mooring to a station in front of the naval Yard "for the protection of that establishment". This move must have been more symbolic than anything else for a week later she was back in her accustomed position. If there was going to be a conflict Hong Kong would not be at the centre of it. Its great value was as a base: on May 7th the cruiser H.M.S. Shannon (ship-rigged, and with a single screw propeller that could be raised at will, showing how slowly, if not reluctantly, the Navy was committing itself to steam), and the composite corvettes H.M. Ships Diamond and Ruby called in to refuel and reprovision, but they were soon off to the north where the action would take place. In June likewise, the steam corvette H.M.S. Charybdis (17 guns and 400 H.P.) arrived, went into dock at Aberdeen, but left for Shanghai after a quick turn-round of less than a fortnight.

Those left behind continued to maintain their supportive role. In the "Daily Press" for May 8th there is what must be one of the earliest accounts of a torpedo trial in the East. The torpedo had been invented by Robert Whitehead, British manager of an engineering firm in the Middle East around 1866. This report states that the two used in this trial were manufactured at the Naval Yard, a tribute to its engineering skill. The account makes interesting reading: "Amongst the defensive measures which have already been taken and are at present being prosecuted for the protection of the Colony is that of the manufacture of torpedoes at the Naval Yard. For some time past a number of these machines have been in the course of construction, between seventy and eighty being now ready for use, and yesterday an experiment was tried in the Ly-ee-moon Pass with the double object of showing

**H.M.S. Audacious, flagship of the China Squadron, at Aberdeen in 1878.**

its effectiveness and giving the men a little torpedo practice. The implement of destruction used was one of the kind known as the outrigger or spar torpedo, and it was dispatched on its errand from an ordinary Service steam pinnace. The charge, which is 100 pounds of powder, is ignited by a frictionable tube—and not by electricity as is the case in most infernal machines—and is contained in a cylindrical wrought iron case. This is lashed on the end of a spar over the bow of the boat, and as soon as she has steamed sufficiently near the object of the attack, the spar is hoved forward by means of blocks and tackle until the torpedo is submerged to an extent of about nine feet, and immediately upon the order "full speed astern" being given by the officer in charge, the firing line is pulled and the explosion takes place.

The junk to be destroyed was towed out in the Pass just this side of Shaujiwan and everything being ready, the pinnace steamed slowly towards her, but it was evident there was some hitch for when the firing line was pulled there was no result; another and another attempt was made, and yet no blow-up. It was then discovered, on more closely examining the drum, that the firing key was broken, and that the water had got in. Luckily, two torpedoes were shipped and immediately the attempt was made with the second one, up went the junk in a thousand pieces and a large column of water at least forty feet high, and although the pinnace went hard astern, she did not clear the radius of the falling water and timbers of the junk soon enough to escape a drenching, and one piece of timber striking Lieutenant Needham on the back, he was knocked overboard and sucked down to some depth, but as he received no injury from the blow, he was soon able to strike out and get on board again.

The experiment was a great success in as much that it showed that a steam launch could emerge from a creek and in a few minutes explode one of these torpedoes under the broadside of an enemy's ship. An ironclad could not withstand such an explosion without being destroyed or totally disabled, and there is little fear of hostile men-of-war being able to enter the harbour with a few torpedo boats always on the "qui vive". The experiment was witnessed by His Excellency the Governor and Commodore Watson, and a large number of naval and military men and public. The arangements were under the supervision of Lieutenant Needham of Victor Emmanuel, as torpedo officer, and concluded about six when quite a flotilla of steam launches returned to Hong Kong." (a banyan with a difference!)

After all this excitement, naval life settled back into routine. The 28th of the month being Queen Victoria's official birthday, all men-of-war in harbour dressed ship, and at noon royal salutes were fired from Victor Emmanuel and the shore battery. It is not clear exactly which battery is referred to in contemporary accounts, but it was probably the one at Wellington Barracks. Nowadays, of course, the Navy keeps its three saluting guns on Bull's Nose Point, beside the entrance to the basin. In 1878 naval ships dressed overall and saluted for George Washington's birthday, for the anniversary of the Queen's accession, for the anniversary of her coronation, for the 4th of July, for the birthday of the Prince of Wales and for the King of Spain's birthday.

And so the hot, sleepy months of summer wore on with the routine of business going on as usual and with leisure hours being filled with country walks, boating trips or sporting activities on the field. There were, of course, other ways of enjoying oneself. In a book written only three years earlier (and there is no reason to doubt that things would have changed by 1878) a visitor to Hong Kong has this to say about a certain locality:

"Passing west along Queen's Road, we came upon a quarter of the town much frequented by seamen of all nationalities. Here, spirits are sold in nearly every second shop and bands of sailors may be seen spending their time and money on questionable drink in more questionable company, roaring out some rough sea-song in drunken chorus, or dancing to the time of a drum and flute, accordian or cornopean. The keepers of these grog-shops might be mistaken for respectable

Home-made entertainment 1878 style. The cast of a production of H.M.S. Pinnafore in the Theatre Royal of the old City Hall. Dick Deadeye looks well cast!

The harbour in 1878 with H.M.S. Victor Emmanuel in centre, the Receiving Ship for the naval establishment from 1875 to 1897. An interesting comparison with a similar modern view on page 64 can be made.

members of society were it not for their bulldog, battered and damaged countenances, which betray sundry evidence of bruises and black eyes, received in taking the change out of their customers. The files of Chinese houses which rise above this locality embrace Tai-Ping-Shan, or the hill of great peace."

In other words, it was the West Point-Kennedy Town area that had the reputation for a run-ashore, rather than Wanchai which, in the nineteenth century, was a fashionable suburb. It is quite possible that in the area mentioned was a certain "Beehive Inn", famous, or infamous, depending on one's viewpoint. Its sign bore this verse:

"Within this hive, we're all alive,
And pleasant is our honey;
If you are dry, step in and try,
We sell(s) for ready money."

Who could resist such an invitation?

Turning to more serious matters, there was still no sign of action against the Russians. By September the ships began to return from the north. The ironclad, H.M.S. Iron Duke, had come out to relieve Audacious as the Admiral's flagship in October. There was the usual leave-taking ceremony. As Audacious steamed slowly through the men-of-war anchorage with Vice Admiral Hillyar on board she was loudly cheered by sailors on the rigging of Victor Emmanuel and the gunboats, while Portuguese, American and Russian ships also joined in. However the German gunboat, Freya, "was conspicuous by her crew remaining below", remarked the "Press" somewhat severely. In the following day's paper, however, appeared an interesting little paragraph. Noting the criticism of the Freya's conduct, the naval authorities had hastened to issue this disclaimer. "So far from her crew remaining below as mentioned, the Freya was the first ship to man her rigging and cheer, while at the same time her band played the National Anthem. In fact, no greater honour could have been paid to the Admiral as the Freya is not a saluting ship. The Audacious acknowledged the compliment by her band striking up the Prussian National Hymn."

Not much was allowed to ruffle the calm waters of possibly the finest harbour in the East!

The year ended very much as it had begun with home-made entertainment onboard Victor Emmanuel and the traditional Victorian Regatta. This time, Commodore Watson laid on some "negro entertainment" for his guests, by courtesy of crew members of Magpie. The programme consisted of minstrelsy and a well-known sketch entitled "The Barber's Shop". (One imagines the latter would bear a strong resemblance to a modern "Crossing the Line" ceremony!) As ever, the evening passed off well, much appreciated by a combined naval and army audience, leavened by the presence of several ladies.

And so the year ended for the naval vessels in harbour: Victor Emmanuel, its despatch vessel, Vigilant, and three gunboats. What had the Navy to show for its year in Hong Kong? No great deeds certainly, no feats of valour or endurance, no dazzling flights of technical brilliance. Why not? The situation simply did not demand such a virtuoso performance: relations with China had remained stable, the threat from Russia did not develop. Instead, the traditional naval virtues had been on display, efficiency, courtesy, camaraderie, and always that joyful willingness to make the best of whatever are the given circumstances.

CHAPTER 4

# EXPANSION THROUGH RECLAMATION 1902 - 1908

The naval authorities, having dug themselves in, so to speak, were still faced with the problem of how to create the space necessary for a modern dockyard. The only possible direction was seawards. Thus it was decided, in the closing year of the nineteenth century, to carry through a massive reclamation scheme before, and also continuous with, the building programme. This naturally entailed an increase in personnel, which can be traced in the Navy Lists for the relevant years. Changes began in 1894 with the appearance of an Assistant Civil Engineer, to be followed in 1896 by a Temporary Superintending Civil Engineer. There was, in fact, nothing very "temporary" about the post as the man concerned, Mr Owen Ordish, did not leave Hong Kong until 1905, and thus truly saw the development programme through from start to finish. At this time too, various officials are listed as having gained assistants, indicating an increase in the volume of work. 1902, the year in which the foundation stone for the extension was laid, sees the list of those involved at its longest. The entry for officers in the dockyard reads:

Naval Officer in Charge
Commander
Chief Constructor & two Assistant
  Constructors (1st Class)
Chief Engineer & three engineers
Naval Stores Officer and Cashier.
Three Assistant Naval Stores Officers
Deputy Expense Accounts Officer
Temporary Superintending Civil Engineer
Temporary Civil Engineer
Assistant Civil Engineer, 1st Grade
Two Temporary Assistant Civil Engineers
Assistant Civil Engineer, 2nd Grade
Boatswain

By 1908 two of the civil engineers had gone home, and by the outbreak of the First World War, there are only two left, because the bulk of the building programme had long since been completed. Of course, as the Civil Engineering Department lessened in importance, so the Mechanical Engineering Department increased in importance. The magnificent facilities created by the one had to be exploited by the other.

The total area finally created was 39½ acres. The new sea walls amounting to 4,580 feet in length, were designed as deep water quay walls, capable of bearing the weight of large guns, boilers and cranes, and of forming the bulwark against the sea. Two novel features were a floating basin and a graving dock. Designed to have an area of 9½ acres and a depth, at the lowest spring tides of 30 feet, the basin was to be square shaped and to be used for the repair and re-fit of vessels while still afloat. Even today, this basin is a well-known landmark, making it easy to identify the position of the naval base along Hong Kong's crowded foreshore. The graving—or dry—dock, with a length of 550 feet, also had a lowest depth of 30 feet. As one of the photogaphs shows, it was capable of taking three submarines at one time.

The carrying through of this massive project undoubtedly posed technical problems. For one thing, much of the work had to be done under water: the walls had to be completed to 4/5ths of their height before they were even visible. This involved the training of about twenty local Chinese divers who were to prove skilful workmen. Additionally, the preparation of the materials was no mean task. The contractors established a huge depot at Mataukok in Kowloon Bay, where they manufactured the concrete blocks with which the sea wall was to be built. This site was chosen because there was an ample supply of granite to hand. Even then, a huge reclamation scheme had to be carried out before the block yard and pier, from which the stone was sent across the harbour, could be built.

A large quantity of plant had also to be provided, much of it specially designed. The

Dockyard in the making . . . The outline of the eastern arm of the basin can be clearly seen.

**The basin is complete and in use in this picture taken in 1914. H.M.S. Tamar is alongside the West Wall. To the right of the photo is Wellington Barracks — taken over in 1946 by the Navy.**

**The main entrance to the Naval Dockyard situated in Queen's Road. After the take-over of the North Barracks, there was another entrance in Murray Road. In 1959, on the surrender of the naval lands, these fine iron gates were offered to the Hong Kong government.**

dredger "St Enoch" was a powerful vessel for those days, capable of carrying 600 tons of soil from a depth of 46 feet. She lifted the hard material from the future basin at the rate of about 1800 tons a day, most of which, of course, was simply moved across the site to the proposed land area. Other machinery consisted of four steam pile-drivers, four locomotives, six portable cranes, thr e ten ton cranes on barges for settling the blocks and numerous wagons and trollies. The graving dock, built behind a coffer dam, required its own pumps and boilers. There were also two pairs of fifteen feet Invincible series centrifugal pumps, driven by vertical compound condensing engines and capable of lifting several thousand gallons per minute from a depth of 62 feet.

Up to the day of the ceremonial laying of the foundation stone, about half of the work of dredging the floating basin, one third of the excavations for the graving dock and four fifths of the foundation work for the quay walls had been completed. The ceremony attracted much public attention at the time and was well written up in the local press. The "China Mail" for the 14th January 1902 makes this interesting comment:

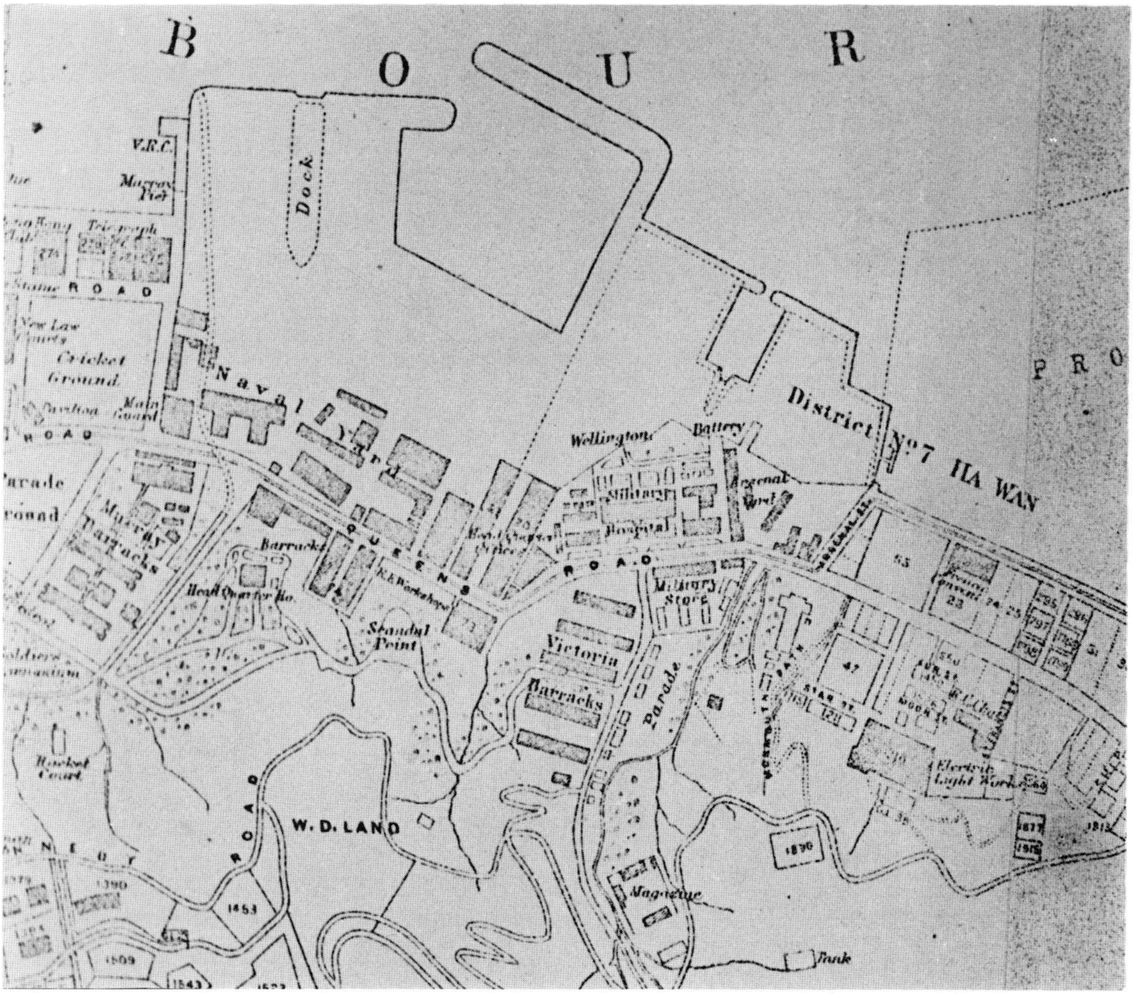


**A plan of Victoria dated 1912. This was the shape of the naval base until 1946 when the Wellington Battery area was absorbed to provide office and living accommodation.**

"It is one of the most obvious proofs that Hong Kong exists primarily as a naval and military base that the wishes of the Colonial Government and the commercial community of the Colony have had the least weight in the decision arrived at respecting the position of the Dockyard and the form the alterations should take."

Times change!

The ceremony did, however, mark the culmination of a great deal of work and planning, and signalled that for the foreseeable future, the Navy would remain at the heart of the rapidly expanding city. Interestingly enough, the memorial stone laid at this time was found in October 1959 when the dry dock was being filled in under the dockyard closure scheme. The "China Mail" for the 1st October that year, states:

"The copy of the China Mail dated January 14th 1902 and of the Hongkong "Daily Press" the day after, six coins ranging from a dollar to a cent and a faded plan showing the extent of the Dockyard extension being undertaken at the time the stone was laid, were in a glass container found under the memorial stone at the head of the dock. The memorial stone and the relics are being retained by the Naval Yard authorities pending Admiralty instructions."

The stone, with its brass inscription, can be seen in a small garden in today's base.

# The Docks in Use

HM Submarines Odin and Otus in dry dock. Measuring 550 feet in length, it had a lowest depth of 30 feet. On the surrender of the dockyard lands in 1957/9, it was filled in, but its entrance, with the depth markings on the wall, can still be seen in HMS Tamar today.

## Between the Wars . . .

Hong Kong expands.... This photograph was probably taken by an aircraft from H.M.S. HERMES seen alongside the North Wall.

An aerial view of the dockyard area — with H.M.S Tamar alongside (centre).

CHAPTER 5

## THE DOCKYARD UNDER ATTACK DECEMBER 1941

The following is an eye-witness account of the Japanese attack on the dockyard, by Commander F.W. Crowther, who was present throughout.

On December 8th at about 5.30 a.m. I was rung up in my house in the dockyard and told the war had started. The Japanese troops had crossed the frontier on the mainland of Hong Kong territory, that is, on the Kowloon side. After informing all concerned by telephone we immediately started everything going according to plan. At 8 a.m. the Japanese sent over about fifty aeroplanes and strafed Kai Tak aerodrome, and all our planes (five, small out-of-date ones) were destroyed, and then we had none at all. Our naval force consisted of three destroyers, two of which were ordered to Singapore, and left that night, so then we had only one, the Thracian. In addition we had one fair sized gunboat available and two small ones, and eight motor torpedo boats. That is all the real fighting Navy we had, but in addition we had some small armed merchant vessels called A.P.V.'s (auxiliary patrol vessels) manned by H.K.R.N.V.R. (Hong Kong Royal Naval Volunteer Reserve) officers with Chinese crews which were used purely for patrol purposes.

So the battle became purely a land affair and most of the naval personnel were employed as soldiers. The total force of our Army including volunteers etc. was 15,500, the Navy including dockyard workmen, volunteers, etc. 960, Air Force about 80. We had about 50 guns, most of them facing seaward, that is, away from the Japanese, and no aeroplanes after the first few hours. The regular army comprised the following regiments: 2nd Royal Scots, 1st Middlesex, 5/7 Rajputs, 2314 Punjabis and two Canadian regiments, the Winnipeg Grenadiers and the Royal Rifles of Canada. The two latter regiments had only recently arrived from garrison duty in Canada and Bermuda, had done no military training and did not know the country locally. I served throughout the battle in the naval dockyard. This dockyard had an overall area of about 600 yards by 300 yards including basin, docks, shops etc., and has a water frontage of about 600 yards. It was not long before we were bombed and occasionally shelled, and when we saw that the Army was unable to hold the mainland, shifted all our stores from the Kowloon dockyard to Aberdeen dockyard on the island, at the same time putting up defences in Hong Kong dockyard.

On the night of the 11th—12th we evacuated our troops and their stores from the mainland to Hong Kong island, the Japanese occupying Kowloon on the 12th. This was much sooner than we expected and we even hoped the Army would be able to hold their line on the hills in the mainland for several weeks. But it was not to be and so on the 12th the Japanese had Kowloon and we could easily see them across the harbour, about 1100 yards away, and sometimes they even potted at us with rifles. They now shelled us in earnest. One morning, in a little over an hour, we counted about 230 shells falling in the dockyard. These shells made a fearful noise both in flight and when they exploded, but did surprisingly little damage. It was, however, extraordinary how few casualties we had when walking about the dockyard. We also had a lot of trouble from the fifth column snipers potting at us from houses in the town behind. Throughout the hours of daylight we had aeroplanes overhead which occasionally dropped bombs on the dockyard, but more often unloaded their cargo on the army positions on the hills and in the town, passing over us on their way.

I did not leave the dockyard as I expected a frontal attack across the harbour at any time, particularly at night. Apart from the defences in the yard we were fully employed putting the fires out which were caused by shells and bombs. We were usually shelled whilst doing

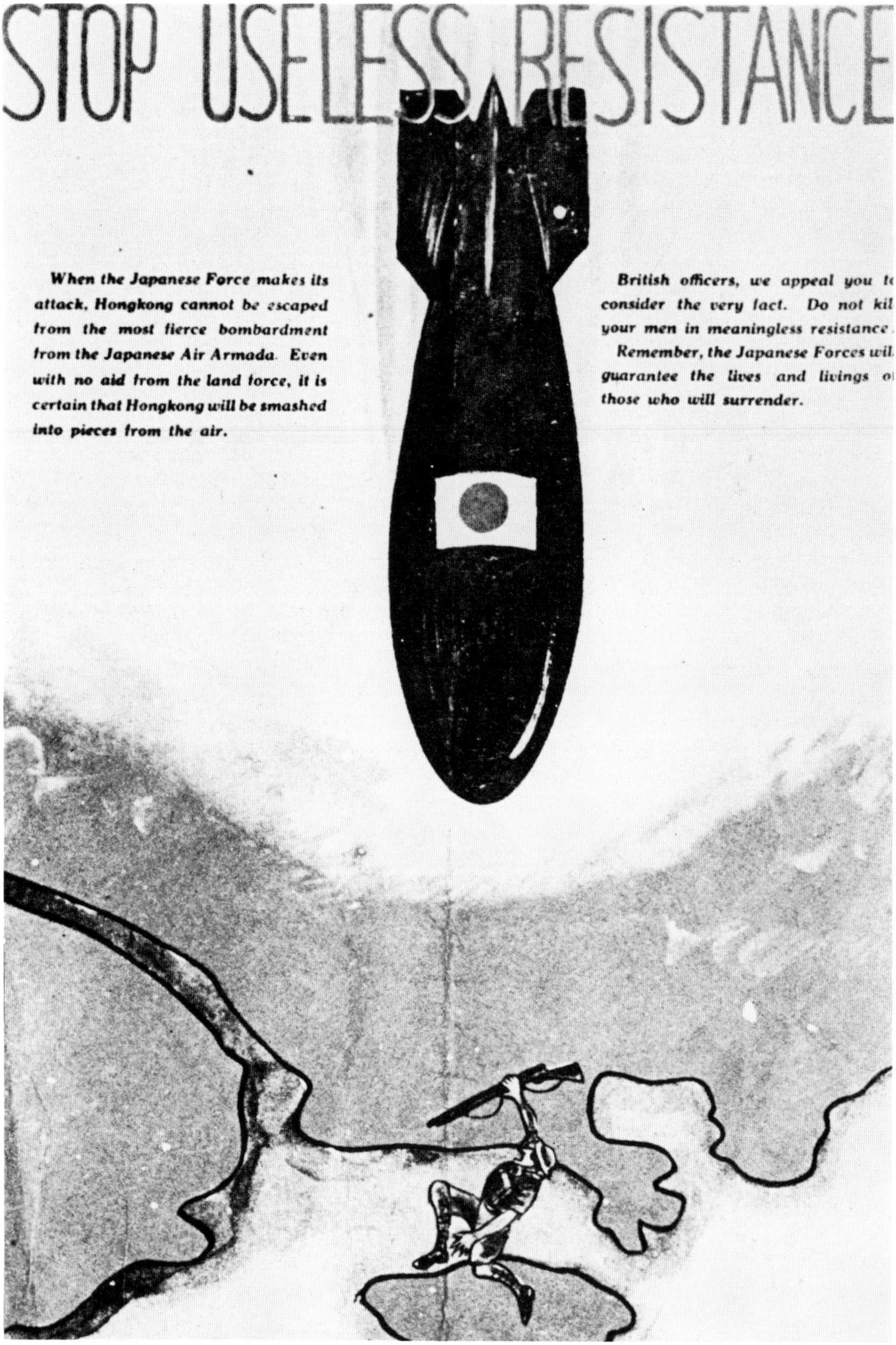
STOP USELESS RESISTANCE
When the Japanese Force makes its attack, Hongkong cannot be escaped from the most fierce bombardment from the Japanese Air Armada. Even with no aid from the land force, it is certain that Hongkong will be smashed into pieces from the air.
British officers, we appeal you t consider the very fact. Do not kil your men in meaningless resistance.
Remember, the Japanese Forces wil guarantee the lives and livings o those who will surrender.

◀ A copy of the leaflet dropped on Hong Kong — prior to invasion.

this, and on one occasion they got a direct hit with a shell on the fire engine, seriously wounding two of our men, both of whom are miraculously still alive and practically all right now. Although I was only about five yards away at the time I did not get a scratch. Several shells fell close to me on many occasions and many just over my head, but I was not touched; there were many others like me, and most who did get hit by a bit of shrapnel did not get much of a wound. The noise, however, was terrific. We only had one or two actually killed. The old Tamar, the wooden depot ship, was taken out of the dockyard on the 8th December and we sank her at her buoy on the 12th. We also destroyed all confidential matter not actually in use on the 12th. On the 13th December the few troops we had on Devil's Peak peninsula were taken to the island of Hong Kong, and then the Japanese had complete control of the mainland which put them only 800 yards off the island.

At 10.30 a.m. on the 13th the Japanese delegation crossed the harbour under a flag of truce and demanded our surrender which was refused.On the 14th the air raid warnings which were a farce anyhow were discontinued as the Japanese were using Kai Tak aerodrome only 3½ miles away from the dockyard and were over all day bombing without cessation. They did not bomb at night however. Our only destroyer, the "Thracian", grounded at Uk Kok when going into harbour at night to attack some junks full of Japanese troops, but although badly damaged, managed to get off quickly and destroy the junks. I sent Lieutenant Cole, R.N. to look after Aberdeen dockyard as I expected the Japanese to attack our dockyard at Hong Kong at any time, which they could have done easily. On the 15th December we were heavily shelled and occasionally bombed in the dockyard, as we were every day throughout, so I will not mention it again. That evening, at sunset, the Thracian was docked at Aberdeen. On the 16th at 13.30 hours there was a very accurate, high level bombing attack on Aberdeen dockyard, apparently after the "Thracian". Big fires were started and there were many casualties. Lieutenant Cole was killed. The Thracian had a near miss, killing a few, but she herself was not seriously damaged. As it was found that the damage suffered by the Thracian when grounding was greater than could be repaired at Aberdeen, all her armament was removed and at 05.30 on the 17th she was undocked and steamed to a nearby island where she was run aground and abandoned. Then we had no destroyer. At about 11.00 on the same day another Japanese delegation crossed the harbour and demanded our surrender. This was again refused but the Japanese seemed surprised and gave us until 16.00 to think it over. This enabled us in the dockyard to improve our defences.

On the 18th after a heavy bombardment the Japanese landed at about 9 p.m. on North Point and Taikoo dockyard and were in the centre of the island by daybreak, having annihilated all opposition. Whenever they were stopped by our troops they simply streamed past on both flanks and surrounded our men. The guns that could bear on the boats crossing the harbour were soon put out of action. Some M.T.B.'s were sent to attack them the next morning: the first two did some damage and sank several craft loaded with troops, but then the Japanese got wise and the next three were immediately sunk. Further M.T.B. attacks were then abandoned. The Army counter attacks failed. All our spare officers and men including the Thracian's crew joined up with the Army and many of them were killed in ambushes. On the 19th December the Japanese were right across the island. The dockyard got more and more shelling and our troops were gradually pushed back. The Japanese put up a battery at North Point and so now in the dockyard we were under a crossfire from Kowloon and North Point. This went on until the 25th December. During daylight hours we were continually putting fires out, building or repairing sand bag posts, getting our food where we could. At night I slept in different parts of the yard but usually with the police near the Main Gate, being a central position, and used to go round all the posts at different times during the night to see that everything was all right. Commander Craven was in charge of the defences and I helped him. Often, and especially towards the end, we had an alarm during the night and stood to: generally it was a report that a boat was crossing the harbour, and we always stood to just before daybreak and just after sunset. But they never came our way. At 8 a.m. on the 25th December I attended Holy Communion in a room above our dispensary with the shells shrieking overhead and occasionally one bursting close by. There were only three there besides the padre. That afternoon at about 3.30 p.m. I was standing to with about twelve sailors by a large shell hole in the dockyard wall on the east side, facing the enemy who were then about 200 yards down the road with nothing between us, as they had broken through. We only had rifles (I had a revolver) but no machine guns or hand grenades, and I was wondering what sort of a show we would put up when I was informed to cease fire. We had surrendered. The dockyard was heavily shelled for several hours after. We had previously blown up all our cranes and pumping plant. Fifth columnist snipers continued to fire odd shots at us in the dockyard, but nobody was hurt. During the night large numbers of men from naval units drifted in from outside where they had been attached to the Army. We slept in the dockyard and in the morning of the 26th the Japanese army marched into the dockyard and took all our arms. The Japanese informed us we were their prisoners and would have to obey their orders. We were then moved from one part of the dockyard to another. The Japanese apparently had no clear idea what they wanted to do with us, but eventually made up their minds that they did not want us in the dockyard. We were then ordered to the China Fleet Club, a few hundred yards down the street. When we got there, carrying all our luggage, other Japanese, apparently very angry, ordered us back to the Yard. So we went back rather exhausted and slept where we could in the Yard that night. They made no attempt to feed us till the evening of December 30th, but we got along as there was food about in dumps, but water was the chief difficulty as the fighting had destroyed the main water supply.

On the 28th we were bundled out of the Yard carying all our luggage and sent to Murray Barracks where most of the Army was. Our Army officers informed us that there was no room for us, but that did not affect the Japanese and eventually we squeezed in somehow after a good deal of trouble. On the 30th at 0700 hours we were fallen in and marched down after several hours waiting, to the ferry. We were then all packed in like sardines into a ferry which was grossly overloaded and taken across the harbour and landed at Kowloon, some three miles from where we were going to, instead of at a pier quite close to our new camp. We were then marched a long way round, up and down concrete streets, carrying all we had, presumably to show us to the local Chinese. The Chinese, however, did not look at all happy and simply looked at us in a dazed silence. The whole road was littered with gear people could no longer carry. I carried a camp bed, blankets, mess utensils, boots and shoes, and as many towels, socks, shorts, vests and uniform suits I could cram in—enough to last me one year. Little did I think that what I carried would have to last me three years eight months with practically no replacements: in fact, most of what I am wearing now (4th October 1945) is what I carried then. I continually dropped my heavy packages and sat in the road until a Japanese came along with a fixed bayonet and kicked me up. Eventually I got a Chinese coolie from the large numbers at the side of the road to carry part of my luggage and then a great number of others did the same, and thus I arrived at Shamshuipo Camp at about 1700 with all my gear including the bed that I started off with, completely dead beat. The Japanese never gave us any food or water all that day, but we got water and a half cup of rice that night, but we were too tired to do

British servicemen being marched off to internment shortly after the fall of the colony. The site is at the foot of Garden Road.

anything. The camp was in a filthy mess, about a foot of dirt and rubble everywhere, no beds or furniture, no windows, and very few doors to the huts. I was one of the few who had a bed, the others just lay down on the dirt for the night. Now we started our Prisoner of War life.

## THE BATTLE FOR HONG KONG

There are many interesting points in this account. The first lies in the opening sentence: for Hong Kong, the war did, indeed, start on December 8th 1941, and not on September 3rd 1939 as it did for the mother country. As during World War One, the Second World War had, until that day in December, largely passed the colony by. The second point is that obviously it had been decided that there would be no serious naval defence of Hong Kong since two of the three available destroyers were ordered to Singapore on the evening of the first day of the Japanese invasion in the New Territories. The line-up of what remained was not impressive: an old destroyer, H.M.S. Thracian, built in 1916, with a maximum speed of 21 knots, had a total armament of three four inch guns. The "fair sized gunboat" mentioned was H.M.S. Cicala, of the Insect Class. In the person of her captain, Lieutenant-Commander Boldero, Cicala had surely the only one-armed captain in the whole of the Royal Navy. Nelson would have approved! Indeed, something of the Nelsonian spirit was desperately needed, and was displayed, during the last naval actions of those sad December days. The final line of naval resistance was represented by the Motor Torpedo Boat Flotilla, rather a grand sounding name for six old craft.

It is amazing, however, what was achieved with such unpromising material. In the fight for Hong Kong, the Navy had several roles to play: to support, by bombardment, army action on land; to evacuate troops; to intercept Japanese landing craft attempting to cross the harbour; and finally, when all else had failed, to see to the destruction of shipping too valuable to be allowed to fall into enemy hands. As Commander Crowther says at the beginning of his second paragraph, it was the Navy which carried out a miniature Dunkirk between the 11th and 13th of December, Thracian being instrumental in evacuating the Rajputs from the Devil's Peak area, coming in as close as her captain, Lieutenant Commander Pears, dared. Even so, shells and mortar bombs exploded round the troops as they waded out to the ships which then took them to Aberdeen, rapidly becoming the centre of resistance on the island, probably because of the simple geographical fact that it was on the coast further away from Japanese guns.

Having established themselves on the mainland, the enemy attempted to cross the harbour on the night of December 15th. Warned by the shore batteries on the north side, Thracian sailed from Aberdeen round into Kowloon Bay, darkness providing sufficient cover against the enemy guns. The need to keep close to the shore, however, necessitated running some risks, and the ship went aground, flooding her forward compartment. Desperate efforts to refloat her succeeded just before dawn broke to reveal a sitting target, if ever there was one: two large junks packed to the gunwales with invasion troops, which Thracian promptly blew out of the water before limping back to dry dock. As Commander Crowther's account shows, the necessary repairs proved too difficult, and two days later she was run aground on a nearby island.

Effective though all this was as a gesture of defiance, it was obvious that more action was needed to intercept the Japanese who, on the night of December 18th, landed along the north coast, between North Point and Lyemun. Accordingly, the S.N.O., Commodore A.C. Collinson, ordered five motor torpedo boats (three of which were commanded by members of the Hong Kong Royal Naval Volunteer Reserve) to the attack. All five having formed up behind Green Island, the first pair moved off. It proved to be the most successful, managing to sink three heavily laden landing craft, enough to make the enemy stop its ferrying service . . . but only temporarily. An alternative which was denied

**H.M.S. CICALA—The last RN vessel to be put out of action in 1941.**

**.... and Her crew, in earlier and happier days.**

to the defenders, namely air power, was brought into play, with the result that the second pair had little chance. Commander Crowther is slightly inaccurate when he says that the remaining three boats were sunk; one of the second pair, through taking evasive action, did manage to escape. The slaughter, however, was such that the action was abandoned, but either through bad communications or the desire to "have a go", the last boat, M.T.B. 26, went ahead, but, sailing right against the enemy guns, was blown to pieces.

The end was in sight. David might annoy Goliath from time to time, but he could not stop him doing what he willed. H.M.S. Cicala, which had been engaged in the melancholy task of sinking merchant ships, was the last to be put out of action, on the 20th December, again through air attack. Her crew, however, managed to take to the Carley floats and four hours later were fighting as soldiers around Bennett's Hill in the west of the island. As Commander Crowther points out, the Navy was increasingly forced to fight on land, as its vessels were destroyed. The Industrial School at Aberdeen became the new headquarters from where naval parties joined the Army in the fight for Stanley and Repulse Bay. A story of the fate of one such expedition seems typical of the time: a party of thirty ratings from H.M.S. Thracian had been sent out to support the hard-pressed Canadians in the direction of the Wongneichong Gap. Their lorries were ambushed on the coast road, however, but four survivors managed to escape to a nearby house. After holing up for several hours they came out, as they thought, under the protection of the white flag of truce. Three were immediately bayoneted but the fourth, Able Seaman Ronald Mattieson, seeing the fate of his mates, struck the rifle out of the hand of the advancing Japanese soldier, and dived headlong over the cliffside, rolling and tumbling the fifty yards down to the sea's edge, where he found a small cave which became his home for the next month. At the price of a broken collar-bone he had saved his life.

Looking back over these events it must be said that the best results were achieved with the very limited resources available in personnel and equipment, but that these resources were pitifully inadequate to meet the task on hand.

CHAPTER 6

## THE RETURN, AUGUST 1945

The first communique from the government to the people of Hong Kong since December 1941 was issued on August 30th 1945. It read:

Rear Admiral Harcourt is lying outside Hong Kong with a very strong fleet. The Naval dockyard is to be ready for his arrival by noon today. Admiral Harcourt will enter the harbour having transferred his flag to the cruiser "Swiftsure" which will be accompanied by destroyers and submarines. The capital ships will follow as soon as a passage has been swept. The fleet includes two aircraft carriers, the "Indomitable" and the "Venerable", the battleship "Anson" and the cruiser "Eurylus".

We are fortunate in having an eye-witness account of the Fleet's entry, written by Commander MacRitchie on board the Canadian anti-aircraft carrier, "Prince Robert".

"On both sides all the way up channel we could see the havoc wrought by British and American planes in earlier months. It would appear that Japanese policy had been not to repair the demolition damage created by the British prior to the fall of Hong Kong, or by the allied planes. With the exception of two old crocks the harbour was destitute of Japanese fighting ships. We proceeded to the Kowloon side while Swiftsure set about dropping anchor close in to Hong Kong dockyard. A Canadian ship would be the first allied warship to go alongside a jetty in the colony since war begun.

Small groups of Chinese, intermingled with whites, watched from without the dock gates as we eased our way in. At the dockyard gate was a Japanese sentry. There was no frantic waving or demonstrations until three platoons of "Prince Robert's" men, in full marching order, complete with fixed bayonets, clashed over the gangplank at the captain's order "Secure". It was then that the crowd broke loose, left the Japanese staring blankly, and surged onto the wharves.

It took our men only a matter of minutes to accomplish the occupation of the Kowloon dockyard area. Across the harbour in Hong Kong the Royal Navy units were enjoying equal success. The Japanese soldiers were speedily rounded up and moved outside the gates to their military quarters across the street. (There were about 1,500 Japanese soldiers in Kowloon dockyard.) There were no untoward incidents."

Across the harbour, the Return was proceeding just as smoothly. The first men to land at the dockyard were from H.M.S. Swiftsure, led by Commander W.L.M. Brown, D.S.C. They were met by a Japanese officer who asked, "Would you mind waiting for the Chief of Staff?" to which Commander Brown replied, "We have no further need of the Japanese Chief of Staff. My orders are to clear the Dockyard of all Japanese." This was done in the first hours after the landing by 500 men from the Swiftsure and Eurylus; 600 Japanese were taken prisoner.

The same day Admiral Harcourt drove to the Stanley Internment Camp for the flag raising ceremony. It was, naturally a most moving occasion, the echo of which remains in the last words of the admiral's speech to the ex-internees, 2,000 of whom were present: "I only want to say one thing. The motive that has inspired all my men, all the blue jackets, to get here as quickly as possible—which we have done—has been, really . . . you people."

On September 2nd Admiral Harcourt was appointed Commander-in-Chief and head of the Military Administration. The task was to return the colony's life to normal as quickly as was practicable. For the time being civilian organisations drew on naval personnel to make up their depleted numbers, naval doctors worked at Stanley while the Harbour Office, with Commander Jolly at its head, was reactivated. The task of policing the colony also fell to the Navy, for which available

The Royal Navy comes back to Hong Kong after three-and-a-half years. The Cruisers SWIFTSURE and EURYALUS in Victoria Harbour.

**Japanese prisoners being marched off to work at the docks.**

personnel were divided into two groups: one became known as the Brown Force, as its head was the Commander Brown already mentioned, while the other was Kennedy Force, in charge of Commander A.R. Kennedy. Men for the two parties were at first drawn in from H.M. Ships Anson, Swiftsure, Eurylus and Indomitable. The island was divided into sections for patrol. Brown Force took over Aberdeen, Taikoo, Stanley and the Peak. When fully developed the two forces comprised about 550 men each, of whom about 450 would be on active patrol, the remainder being occupied on administrative, medical and transport duties. The Hong Kong Police Force was not, at that time, strong enough to act independently. The parties spent some time scouring the ammunition dumps left by the Japanese all over the island, much of it captured in 1941, looking for blanks with which to fire the twenty-one gun salute when the surrender

**Some of the first civilian internees to be released from Stanley Camp. In the early days of re-occupation the Navy did much to help life to get back to normal.**

was signed, as warships have no room for blank charges in wartime.

At Kai Tak airstrip MONAB 8 (Mobile Operation Naval Air Base) had set up shop. The idea of MONABS was born, in the latter years of the war, because of the vastness of the Pacific Ocean and the difficulty of over-flying it, with the consequent fatigue to air-crews based on carriers with limited rest and recreational facilities. The plan was that when an island was retaken, the American Construction Battalions would move in, bulldoze an airstrip and erect accommodation at which point a Monab would take over, providing a base for the squadron to be rested and the aircraft repaired. The process was to be repeated nearer and nearer to Japan. The dropping of the atomic bomb, of course, put an end to such plans as being no longer necessary, which may be the reason why only Monab 8 seems ever to have been operational.

Thus it was that the Mobile Operation Naval Air Base, commissioned H.M.S. Nab-catcher, arrived in Hong Kong, a fully mobile unit of around 700 strong, complete with its own Bedford 4x4's, jeeps, tents, kitchen, craft repair equipment plus a full complement of specialist personnel for aircraft maintenance. It had arrived on board H.M.S. Slinger from Australia where it had loaded up and where it had left MONABS 1—7.

The Corsairs, Barracudas and the one Avenger of the unit, in all about 25 aircraft, flew dawn and dusk patrols right through the last months of 1945. A radar station was also set up in the hills of the New Territories. The Royal Air Force was also based at Kai Tak on the far side of the airstrip. The MONAB unit

H.M.S. SLINGER — which had brought the MONAB from Australia — ferries stores ashore.

Lightering a Fleet Air Arm Barracuda ashore.

The MONAB at Kai Tak airfield.

When Rear Admiral C.S. Daniel took command of the Task Group he flew his flag in the battleship H.M.S. ANSON seen here at anchor in Hong Kong Harbour.

**Japanese prisoners of war bow to the white ensign—shortly after Hong Kong had been re-established as a British Colony.**

found daily work for about 2000 of the Japanese prisoners of war housed at Sham-Shui-Po, the camp previously occupied by Allied troops. Many, employed in cleaning the aircraft, digging and general tasks were still there at Christmas 1945.

More important even for morale was the way the Service brought a welcome return to normality in the social scene: a children's party was held on board Swiftsure, men from H.M.S. Vengeance were behind the re-opening of the Yacht Club on October 2nd, while earlier, on September 10th, the South China Morning Post had reported joyfully:
"A sure sign of Hong Kong's gradual return to normal conditions was the sight of an energetic number of white figures running about on the Cricket Club's ground yesterday, when two naval teams were engaged in a normal match."
Apparently, the scoreboard had long since disappeared for firewood, and the numbers were rusting in the pavilion store-room . . . but the game still went on!

Nor did the evenings drag. Once again, the newspaper's correspondent was on the spot at a dance in Stanley to record that "great enthusiasm was shown by the ex-ternees over two dances, the "Hokey-Kokey" and the "Conga". These dances are very popular in Great Britain and America, but naturally, new to Stanley folks. Several of the Stanley ladies are developing into first class "jitterbugs" under the expert tuition of the young, rhythm-minded members of the Fleet."

The natural culmination to this light-heartedness was Victory Day, October 9th, when "bright sunshine, happy crowds, brilliant uniforms, martial music and flags" formed the background to the official celebrations. The march past, held outside the Hong Kong and Shanghai Bank, took approximately fifteen minutes.

It is difficult at this distance in time to capture the feelings of those for whom an ordeal had just ended, but a letter published not long after the arrival of the Fleet seems to express the feelings of the man-in-the-street. It reads:

"Polite, neat, kind and forebearing, the members of the Relief Staff are carrying out their duty to the satisfaction of all people. We see more sympathy with the poor in the streets than show of pride in being a victorious force. The admiral, commanders and seamen all alike put on smiling faces wherever they be met with. Despite having fought for years, as some officers of the Swiftsure said to me, their calm behaviour and quiet personality still exist. This is the credit of the British Navy."

One can certainly imagine the great sense of relief, that a Service, much loved and blessedly familiar, had finally taken control.

**Admiral Harcourt reads the surrender terms to the former Japanese base commander.**

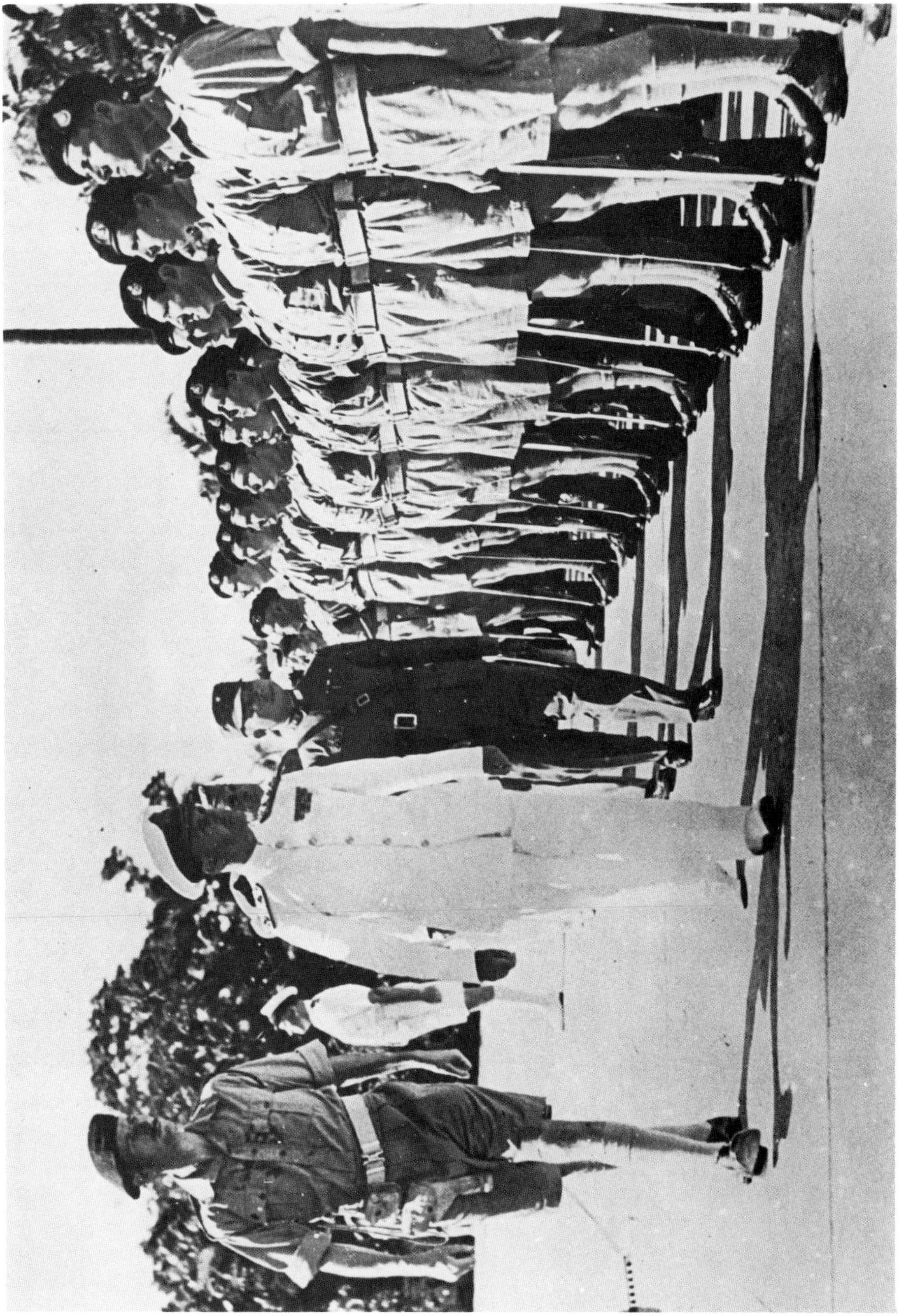

Royal Marines of the liberating force are inspected by Admiral Harcourt at Government House.

**Sailors and marines present arms as the Union Jack is raised again at Government House.**

9 October 1945 — The Victory Parade showing the saluting stand at the entrance to the Hong Kong and Shanghai Bank, with the cricket club in the distance.

CHAPTER 7

# THE CLOSURE OF THE DOCKYARD 1957-1959

On November 28th 1957 it was announced that the naval dockyard in Hong Kong would be closed down. In giving the reasons for this decision, the statement made in Parliament at the time referred to "the reorganisation of H.M. naval forces to meet modern conditions" and to the fact that "future requirements of H.M. Ships in Far Eastern waters will no longer justify the maintenance of a full-scale refitting dockyard in Hong Kong." Economic considerations were also of great importance: it was estimated that the home government would save about £1,500,000 a year through the closure. The decision, however, had been taken with very real regret. "The long association of the dockyard with Hong Kong (fifty-five years from the foundation stone ceremony) and the loss of employment for many workers which must inevitably result from the closure, are factors which have weighed heavily with Her Majesty's Ministers." To soften the blow as much as possible, it was decided that the Yard should be run down over a period of two years to allow for a phased redundancy programme. In 1957 the Navy employed about 4,500 workmen in the dockyard, possessing varying degrees of skill. There were also about 250 Europeans in managerial positions and about 250 clerical workers.

Although the decision was received in the colony with concern, there was little formal protest, due to excellent liaison between the Navy and the government over resettlement and the early reassurances given that everything possible would be done to protect the livelihood of those affected. The run-down started at the beginning of January 1958 with the dismissal of twenty-eight employees from various departments. It was planned that the average rate of discharge would work out at about 130 men a month, of whom some would be able to find work before their dismissal, while others would need the help of the newly formed Employment Liaison Office. At the invitation of the Commodore, employers were encouraged to visit the dockyard to find out more about the type of work done. All went well until the beginning of March 1958 when workers staged a squat-down protest against more dismissals. It all ended peacefully, however, a few hours later and the fact that there was no repetition was undoubtedly due to the strenuous—and in the main successful efforts that both service and civil authorities made to find jobs for the unemployed as quickly as possible, In the first year of the run-down, 2,297 men were released, about half the workforce; by the closure date, November 28th 1959, 4,650 had gone, of whom one sixth could not immediately be found employment.

The employment question was not, however, the sole issue raised by the closure of the dockyard. For the Hong Kong government it meant the realisation of long hoped for plans, that is, the removal of the Queen's Road bottleneck by the acquisition of this considerable area of land in Central, through which one, or possibly two roads could be built to link with Wanchai.

The map illustrates the area involved and the changes planned. Most of the released land had been naval but the section released by the Army was to the east, next to the Police Headquarters. It had been acquired in 1946 by the Navy to provide administrative and living accommodation lost when H.M.S. Tamar was scuttled in 1941. The Hong Kong government paid the War Department 24 million dollars (H.K.) in 1959 for this area. It was not, in fact, handed over until 1962 although the Army agreed to surrender any land necessary for the road at six months' notice, since the new road, to be named Harcourt Road after the admiral commanding the liberation fleet, and which would link Connaught Road with Gloucester Road had top priority in the government view. The delay in handing over the Tamar land was neces-

**1958 — Workers "Squat down" in the dockyard—in protest against dismissals. Outside the gates they are supported by their families.**

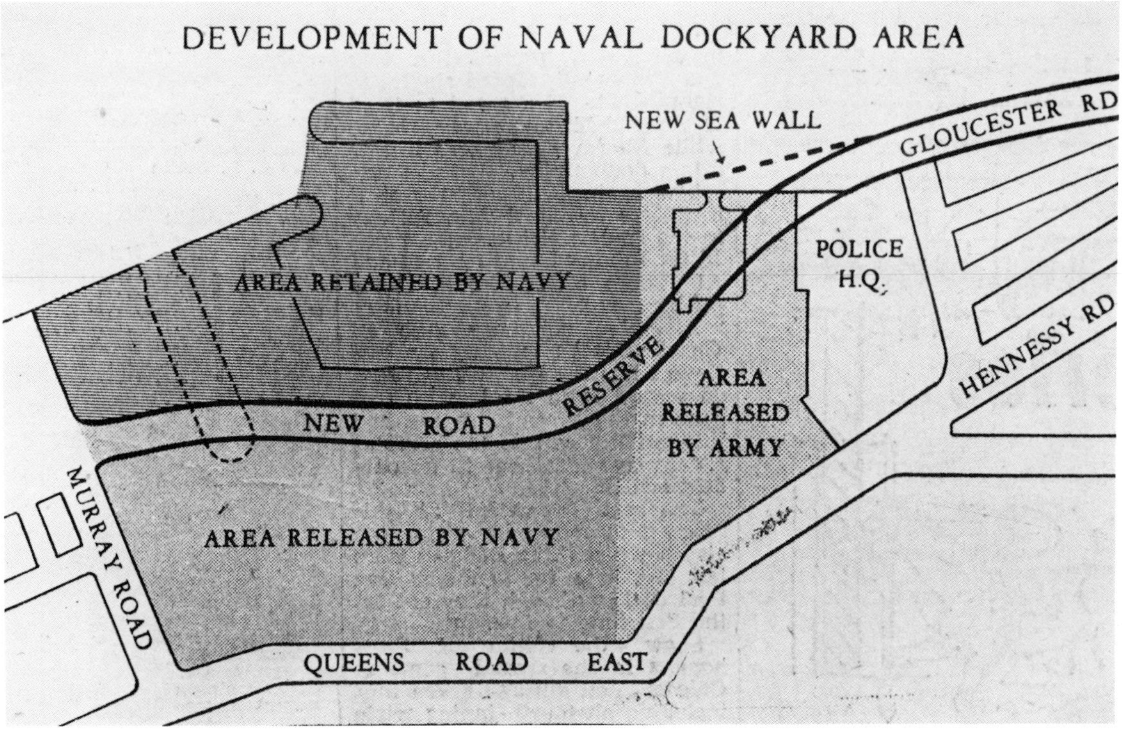


**Re-Development of Naval Dockyard Area 1957/59.**

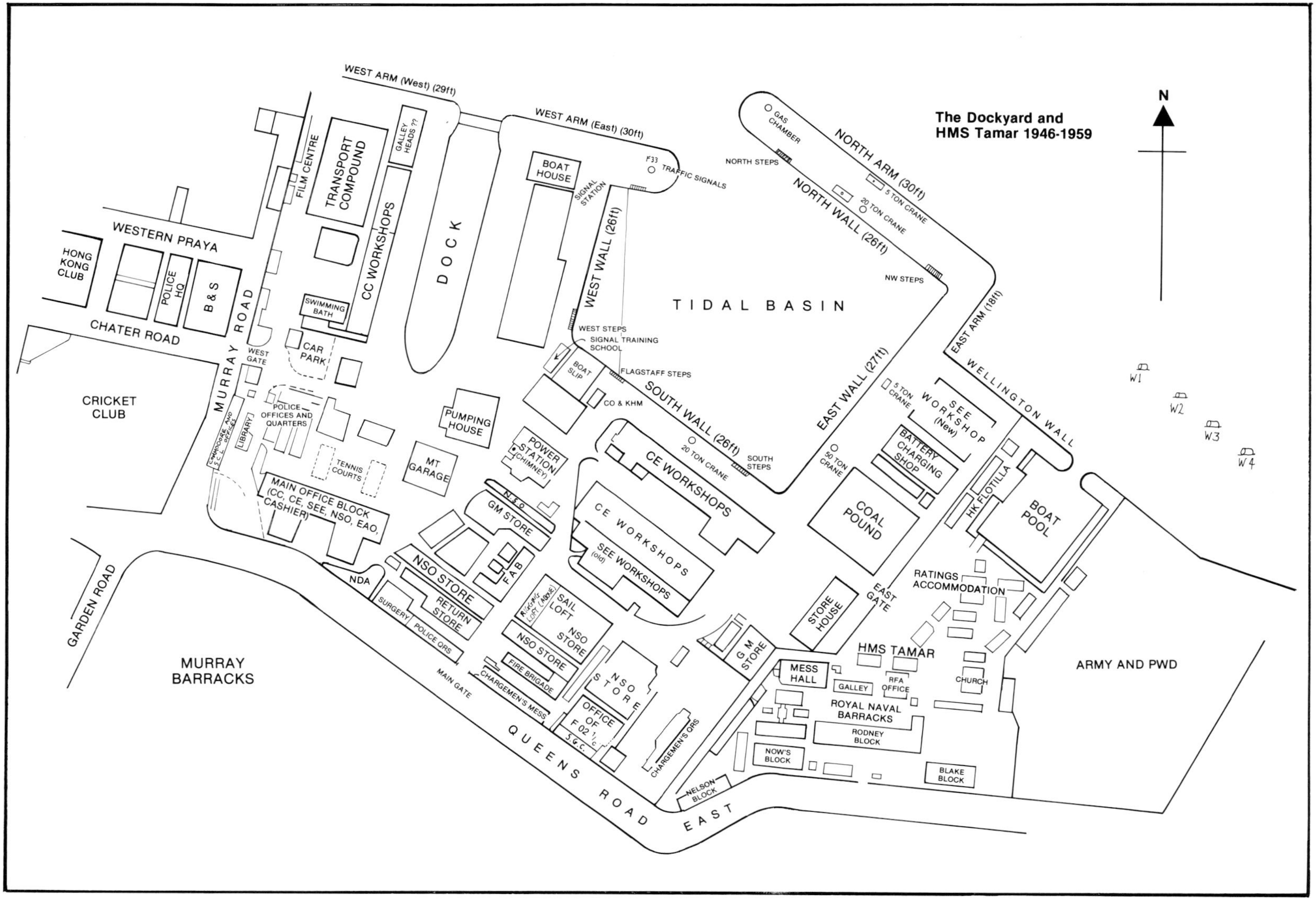


The Dockyard & HMS TAMAR 1946/59

sitated by reprovisioning. Although the dockyard was being rundown, quite a considerable naval establishment would remain there and it all had to be reprovisioned within the restricted area the Admiratly would retain.

Initially the disposal of the land was the subject of negotiation between the government and the Service ministries. Although it was always intended that the Navy retain "a small maintenance and operation base", it is interesting to note that its present site, that is, a corner of what it once had, was not immediately fixed upon. There was talk of "looking around" for a suitable place to establish a small repair yard, and it was not until August 1959 that the public learned, with some surprise, that the Navy had retained a toe-hold in Central, and a sizeable portion of its old waterfront. The rest was handed over to the Hong Kong government on November 28th 1959. The same day the Commander-in-Charge received a farewell message from the Civil Lords of the Admiralty:

"The closing of Hong Kong Dockyard ends a long chapter of honourable service to the Royal Navy in Hong Kong for well over half a century. In that time generations of Dockyard workers created and upheld a great tradition of quick work, versatile skill and devoted loyalty. Their reputation for these qualitites was known and praised all round the world. Their Lordships desire to express their gratitude for all the benefits thus bestowed upon the Royal Navy. Their deep regret that the Yard had to be closed is mingled with pride in the historic achievements of its workers."

Coast Watching Stations were set up at various points around the periphery of Hong Kong round about 1950. This was to maintain a survey on shipping activity in the main lanes around the colony.

A tragic incident which occurred in 1953 was the attack on a patrolling motor launch, ML 1323, by a Chinese Communist vessel. Normally, these patrols would pass each other with a wave of greeting but this time, to the complete surprise of everyone in the motor launch, the Chinese opened fire and so all personnel took shelter in the forward Bridge house which was lightly armoured. No attempt was made to return fire, but instead to get out of range. Before this could be achieved, a four inch shell entered the Bridge house where it exploded, killing two of the personnel aboard and mortally wounding the commanding officer, a young lieutenant. By cool bravery the Leading Stoker in the engine room kept one engine going at excess speed and went on deck to engage the second steering position at the stern, as the Bridge had been destroyed. With the help of the survivors, one of whom took the wheel, the launch was taken back alongside the jetty at Tai—O. The incident was widely reported at the time both locally and internationally.

CHAPTER 8

## THE FIFTH HMS TAMAR

The new H.M.S. Tamar, and the fifth in line, originally had its being in Wellington Barracks, which the Army had transferred to the Navy in 1946. This suitable site, which backed onto the waterfront, was adjacent to the dockyard on its eastern boundary. Between 1946 and 1959 these two units, the dockyard and the administrative base called H.M.S. Tamar existed side by side but completely independently. The map gives an indication of what was there. On the surrender of the naval lands in 1959 a new site had been found for this establishment which emerged gradually between 1959 and 1962 in the form of a compact, modern naval base between Harcourt Road and the waterfront. This road, designed to link Connaught Road with Gloucester Road and completed in 1961, was built across former dockyard land. On either side of it old naval buildings were pulled down and some were used to help fill in the dry dock. The buildings of the Tamar section were left to the last but by 1962, as a plaque in the present base testifies, the new accommodation and quarters were ready in the western end of the old dockyard and the transfer was made across the several hundred yards from the eastern end. A reasonable idea of what the base looked like then, between 1962 and the building of the new tower block in 1978 can be gained from the photographic section. Its task includes the provision of support facilities for the minesweepers formerly based in Hong Kong and for one frigate or destroyer carrying out self-maintenance, and latterly for the five patrol craft which succeeded them. The additions to the naval force—the Fast Training Boat, H.M.S. Scimitar, the Tug, Clare, and two Hovercrafts—which were brought out to strengthen patrols against illegal immigrants—were all manned, repaired and victualled from the base.

**Divisions on the Bull's Nose with two of the patrol craft in attendance. The entrance to the old dry dock can be clearly seen at the stern of the right-hand patrol craft.**

Harcourt Road, seen in the centre of this photograph, was built across former naval land in 1961. The Nissen huts in the right foreground, together with Blake and Rodney Blocks, were all part of the old HMS Tamar.

A northerly view of "A" Block showing the upper floors of cabin accommodation over HMS Tamar administration offices and Wardroom. CAPIC's offices are on the 1st floor. The saluting guns are in the foreground.

**Even a swimming pool in the centre of town . . . just yards from one of Hong Kong's busiest roads. In the middle distance is the Murray Road Car Park, built on what was once the south-west corner of the dockyard.**

"B" Block, originally built as UK Ratings accommodation, but since 1979 has been used for the same purpose by Locally Entered Personnel.

**Last of the old dockyard buildings . . . where the new tower block now stands. The old MFV in the foreground was used as a ferry to Stonecutters Island W/T station, Married Quarters & Ammunition depot.**

## Always a popular Port of Call—Right in the centre of town too!

HMS Lowestoft in the Basin. Compare the skyline with the picture on page 25.

The Australians found Hong Kong a good "run ashore" port too! Here, HMAS Melbourne arrives for a few days.

With less and less HM ships visiting Hong Kong the tug Clare is pressed into service to give local youngsters a taste of sea air. She was later used in the hunt for illegal immigrants. Building work on the new Tamar is underway in the background.

Looking across the basin to the old range of workshops along the west wall. Note, at the end of the north arm, (R) the white Japanese pill-box, later knocked down to make room for a helicopter landing pad.

The only entrance to HMS Tamar during the 60s & 70s was this one gate. This guard house was demolished to make way for the larger entrance gates in use today.

Boxes are unloaded on the parade ground prior to the erection of a temporary warehouse. In the foreground preliminary work on the new Headquarters British Forces tower block has just started.

CHAPTER 9

## THE HONG KONG CHINESE LOCALLY ENTERED PERSONNEL DIVISION

Chinese living in Hong Kong have served in the ships of the Royal Navy from the nineteenth century, although it was not until 1905 that the Locally Entered Personnel Division was formed and formally became part of the Royal Navy. Numbers in the Division have fluctuated over the years in response to the political situation, from the sixty-eight of 1905 to the two thousand plus in the reformed Division in 1946. The present complement is about three hundred, serving in a wide range of capacities as cooks, stewards, seamen, electrical and engineering ratings, medical and stores assistants. The initial engagement is for three years, renewable for a further twenty-two. Training is provided in H.M.S. Tamar and afloat. The Division has been accustomed to serving at sea: in 1905 it was on armed launches and the gunboats of the Yangtse River, in the 1950's it was during the Korean War and the Confrontation with Malaysia. The "mixed manning" of the present day Hong Kong Squadron began in the 1970's, on an approximate 50-50 basis. Many acted as interpreters on the patrol craft which intercepted the armada of small boats from Vietnam and mainland China when the illegal immigrant problem was at its height.

### JENNY

One sailor on board H.M.S. Dorsetshire recalls that his first memory of Hong Kong in 1933 was of a young Chinese woman climbing up the ship's side via the chains to where he was engaged in taking soundings while going up harbour at about ten knots. She wanted the contract for looking after the ships sides' paintwork—and got it! Whoever this was, the side party girls, working from their sampans, have always been a much loved feature of Hong Kong naval life.

To the thousands of sailors who have served there over the last fifty years, one in

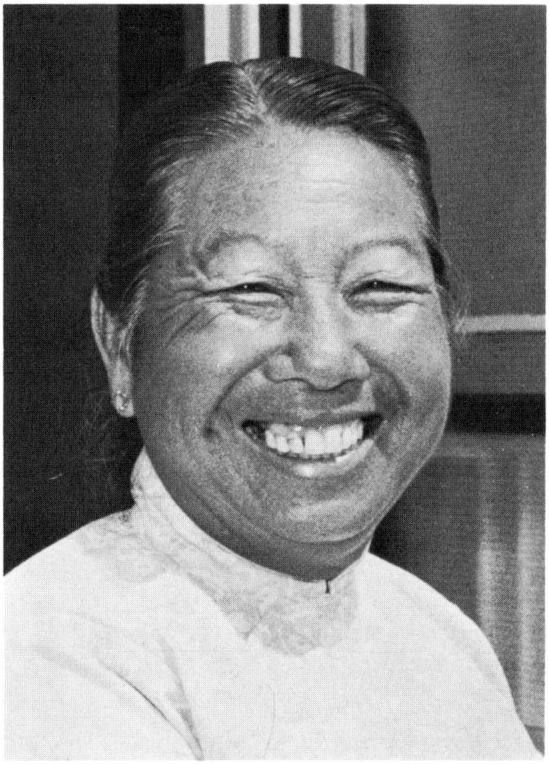

particular, Jenny Sideparty is a well known and well loved figure. Her primary task of leading her sideparty of girls has been greatly augmented by countless acts of generosity and kindness towards so many people, from the most senior admiral to the most recently joined recruit. Nothing is too much trouble, everything is done with that inimitable and infectious smile. Her large collection of photographs is only rivalled by her extensive knowledge of local naval history and personalities. In her traditional costume of baggy black trousers and white, high-collared smock, long pig-tail hanging down the back, she is a link with the past and a reminder of the happy association enjoyed by the Royal Navy with Hong Kong for so many years.

The "Rival Firm"—Susie's sideparty at work on the ships side of HMS WASPERTON.

## CHINA FLEET CLUB

One place which has been visited by all naval personnel, whether on a long or short tour, on a visiting ship or a quick stop-over, is the China Fleet Club. Its foundation stone was laid in 1933 by the then Commander-in-Chief China Station, Admiral Sir Howard Kelly. The building rose on its Wanchai site, at the east end of the dockyard. At that time it faced onto the waterfront but has long since been left inland by reclamation.

In 1941 the club temporarily became one of the centres of resistance to the invading Japanese forces and on the victory of these forces, was briefly used as a dentention centre. After the reoccupation in August 1945 the Club was very quickly re-opened, certainly before the end of the year. With around 25,000 naval personnel in Hong Kong, it was a good time for business. Fortunately the fabric of the building had been little damaged during the occupation and the fighting which preceded it.

The position was, and is, a most valuable one, perhaps too valuable for a six storey building to occupy. In 1981 it was pulled down to make way for a modern, multi-storey block, several floors of which will be the new home of the club. The rest will be let out as offices. Although a well-known landmark has thus disappeared it is in the tradition of the Club to modernise and update its facilities. In 1954, for instance, the Coronation Annexe was completed and an agreement was reached for the United States Navy to open a Purchasing Branch in the building. 1970 saw a new look to the restaurants and family room. The new building, due for completion in 1985, will continue to provide excellent facilities for Royal Naval personnel, by whom it is owned and for whom it is managed.

**The new look China Fleet Club — on the old site. Fleet House was opened on 30th May 1985. The Club only occupies a small proportion of the new building.**

CHAPTER 10

## HEALTH

Disease, rather than the Chinese, was the enemy facing troops and civilians as they struggled to establish themselves on Hong Kong island. Much information on this point is found in Captain Bernard's book, "Narrative of the Voyages and Services of the Nemesis" (published in 1844). This ship, incidentally, made a twin contribution to Hong Kong: its First Officer was Lieutenant Pedder, who became the colony's first Harbour Master, as already described, while its surgeon, Dr Young, having also left the ship, started the Hong Kong Dispensary, better known nowadays as "Watson's, the chemist". The book had this to say about general health conditions:

"Sickness had already begun to prevail among our troops before they had reached Hong Kong (January 1841). The eight days' exposure which they had endured upon the heights of Canton sowed the seeds of ague and dysentry which proved far more formidable enemies than any troops the Chinese could bring against us. After the lapse of a few days, and when the excitement of active operations on shore, and the cheering influence of hope and novelty, had subsided, the sickness spread among the men with alarming rapidity, so that, at length, out of our small force, no less then 1100 men were upon the sick list at Hong Kong. Part of this alarming state of things must be attributed certainly to the pernicious influence of the atmosphere of Hong Kong itself at that season of the year. But every allowance must be made for the exposure which the men had undergone at Canton, and for the susceptibility of constitution produced by long confinement on board ship. The germs of disease were planted in their bodies before the men returned to the harbour of Hong Kong, and therefore an undue stress was laid at the time upon the unhealthiness of Hong Kong itself." Be that as it may, an early blow was struck to naval morale when the Senior Naval Officer, Captain le Fleming Senhouse, fell victim to the prevailing sickness. At his own request, he was taken to Macau for burial (where his grave can be seen today in the Protestant cemetary there), an interesting point in that it reveals the general feeling, mentioned by Captain Bernard, that a re-occupation by the Chinese of Hong Kong island was not out of the question.

Among the first buildings to be constructed on the island were naval and military hospitals, on the site which later became Wellington Barracks, just to the east of the naval yard. The structures were probably of the matshed type, very quick to put up, but equally easy to knock down, as is revealed in an old chronicle which, talking of the typhoon of July 22nd 1841 states, "The over-crowded and badly built hospitals were all levelled to the ground and their fragments whirled through the air." How disheartening it must have been for the colonists of only a few months' standing to begin again and literally "pick up the pieces". One good result, however, was that from then on only substantial buildings were put up, as is shown in early photographs, where the city has a remarkably solid air. Living in the typhoon zone, the Victorian ideal of "building to last" took on a new meaning.

What was probably the first Naval Court of Inquiry to be held in Hong Kong—on April 25th 1841—was concerned with the high death rate and the measures which could be taken to reduce it. The problem was, of couse, that medical science was not sufficiently advanced to provide the correct answers. It is likely that most people were dying from malaria, transmitted through bites by the Anopheles species of mosquito. The resulting fever, though easily seen, was not so easily treated with the medicines of the day. The cause of the fever, too, was wrongly attributed to unhealthy vapours coming through cracks in the ground. As a result, every effort

**The first Royal Naval Hospital at Wanchai, in use from 1873 to 1941. On the site now stands the Ruttonjee Sanatorium, at the point where Kennedy Road meets Queen's Road East.**

was made to find a "healthy spot": the army barracks were moved from West Point to the central district after a disastrous epidemic (General d'Aguilar, the first Commander British Forces, commented in 1845, "to retain Hong Kong will require the loss of a whole regiment every three weeks, and to have 700 effective men, it is necessary to maintain 1400"), while for much of the 40's sailors were discouraged from going ashore too frequently in case they breathed "bad" air. The Navy had few, if any, quarters on land in the early days, though whether conditions were any better on the hulks moored in the harbour is a debatable point, although it was certainly believed to be so at the time. In 1843, when fever killed 24% of troops and 10% of European civilians between May and October, all the European troops were moved out of their barracks and onto ships.

The Navy, of course, looked after its own sick. A Surgeon Woosnam is mentioned in the records for March 1842 because he sat on a Land Committee (which was to portion out the new colony) and another surgeon's name crops up in Janaury 1846, both as isolated instances, but indicating the prescence of a trained practitioner in charge of his own team. In April 1856 a Deputy Inspector of Hospitals and Fleets was appointed (the equivalent of today's Surgeon Commander) to head the growing number of medical officers., They would have worked ashore, possibly at d'Aguilar hospital where operations would have been carried out, and on the hulks moored in the harbour. H.M.S. Minden was the hospital ship which accompanied the fleet sent out from home. Later in the 40's she was relieved by H.M.S. Alligator, which in the 60's was augmented by H.M.S. Melville. A second hospital ship would have been required because after the Second China War (1857-59) the home authorities decided to increase the forces to be stationed permanently at Hong Kong, and consequently the "back-up force" had to be strengthened as well.

Naval doctors were always "on call" for the colony as a whole, to help out in any particularly disastrous situation. Such was the case during the great Bubonic Plague of 1894 which killed thousands. To cope, a permanent Committee of the Sanitary Board was set up, assisted by a naval surgeon, Dr Penny, and two others. Because the existing hospitals were overflowing, special ones were established on board the 'Hygeia" and at a Glass Works Factory, recently completed but not fitted out. Here Chinese doctors worked under the supervision of Surgeon Penny. To make a difficult situation worse, the Europeans were faced with distrust of western medicine. This led many Chinese to conceal their sick and dying, which only served to spread the infection even more. The naval and military authorities, under Commodore Boyes and Major General Barker were asked to start house-to-house searches, which, although effective from the health point of view, aroused great resentment, to the extent of public protests being organised and placards appearing in the streets. Seeking to calm down a somewhat dangerous situation, the Governor, Sir William Robinson, asked the Commodore to station a gunboat opposite the Tung Wah Hospital and Taipingshan. H.M.S. Tweed was accordingly moved into position with the hoped for salutory effect. It must have been a miserable time: admissions ran at sixty to seventy a day, and the Glassworks Hospital had to accomodate two hundred patients in a building suited to half that amount. When it was all over a great parade was held to express the thanks of the colony to both Services.

It was probably the increased naval forces in the 60's which made the medical branch start looking around for its own hospital. It cannot have been wholly satisfactory to treat patients in mid-channel, swinging at anchor, and one wonders what happened during a typhoon. Money was almost certainly a problem. In all probability the Admiralty told the local naval authorities that if they wanted their own hospital they would have to find the resources for it. This would explain the sale of H.M.S. Melville in 1873 for a price of thirty five thousand dollars. With the money was bought the Seamen's Hospital in Wanchai, on a site where the Ruttonjee Memorial Hospital now stands, at the point where Kennedy Road meets Queen's Road East. Although running at a loss when the Navy took it over, the Seamen's Hospital had provided many years of useful service to the community. Opened in 1843 under the charge of Dr Young (the founder of Watson's) and backed by Jardine Matheson and Company, the hospital "flourished", and it was there in 1848 that the first operation in the colony under chloroform was performed by a Dr Harland. At this time in its history the Wanchai district round the hospital was not the congested, noisy area it is today. Indeed, it was a fashionable quarter: just to the west of the peninsula on which the hospital stood was the residential district of Spring Gardens, an idea of which can be gained from a lithograph by W. Bruce, one of a series, which is very popular nowadays with visitors as a souvenir. The print illustrated here shows how close to the sea the hospital was in its early days (patients used to arrive by boat) but reclamation has since left the site far inland. The twin hills on which the hospital and a later adjunct, the Infectious Diseases Hospital, were built, were named Mount Shadwell and Mount Parish after, respectively, the Vice-Admiral and

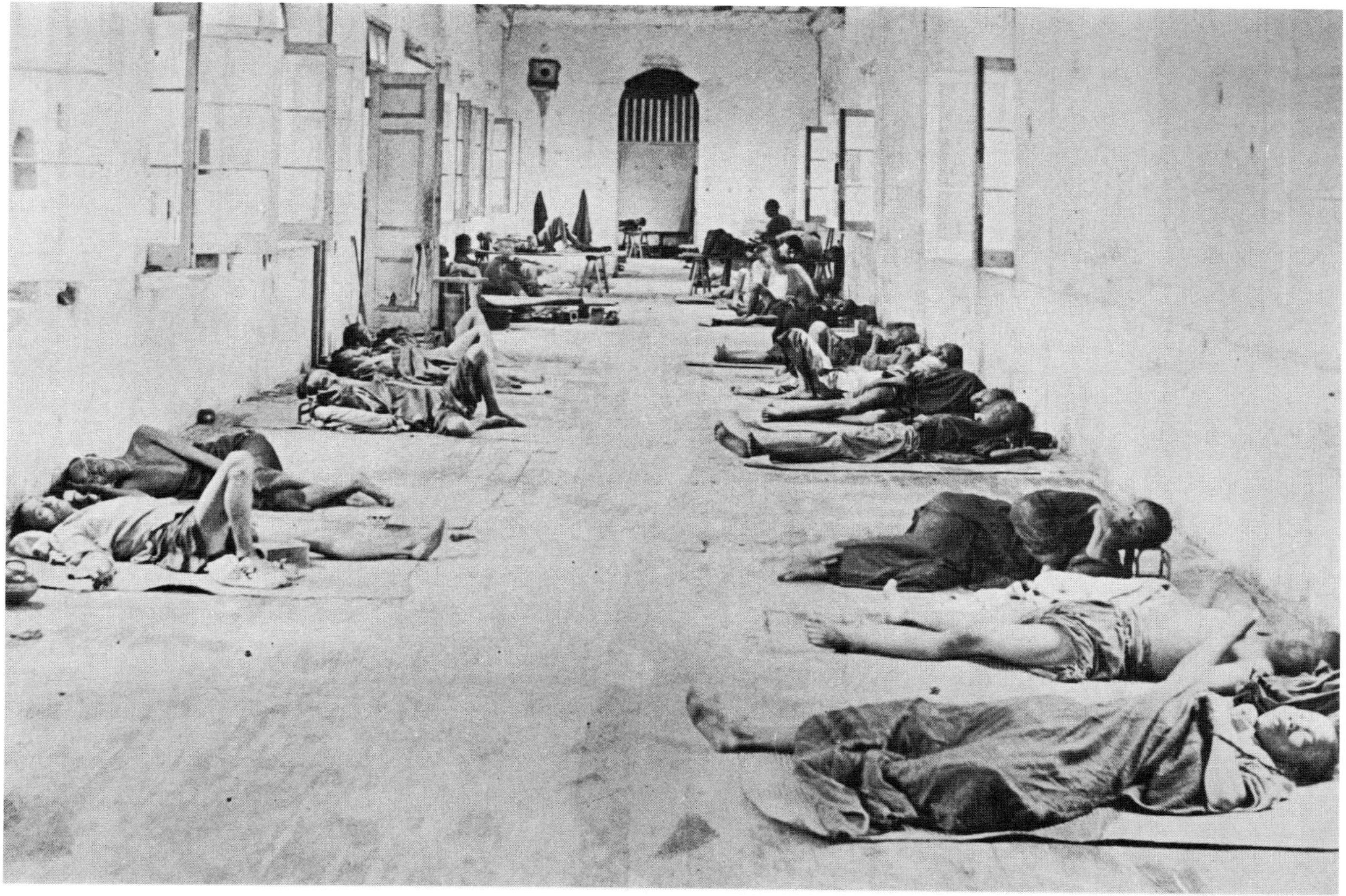

Plague patients in a temporary hospital set up in a glass works factory in Kennedy Town in 1894. A naval surgeon was in charge.

The Second Royal Naval Hospital on the Peak. It was in use from 1949/59.

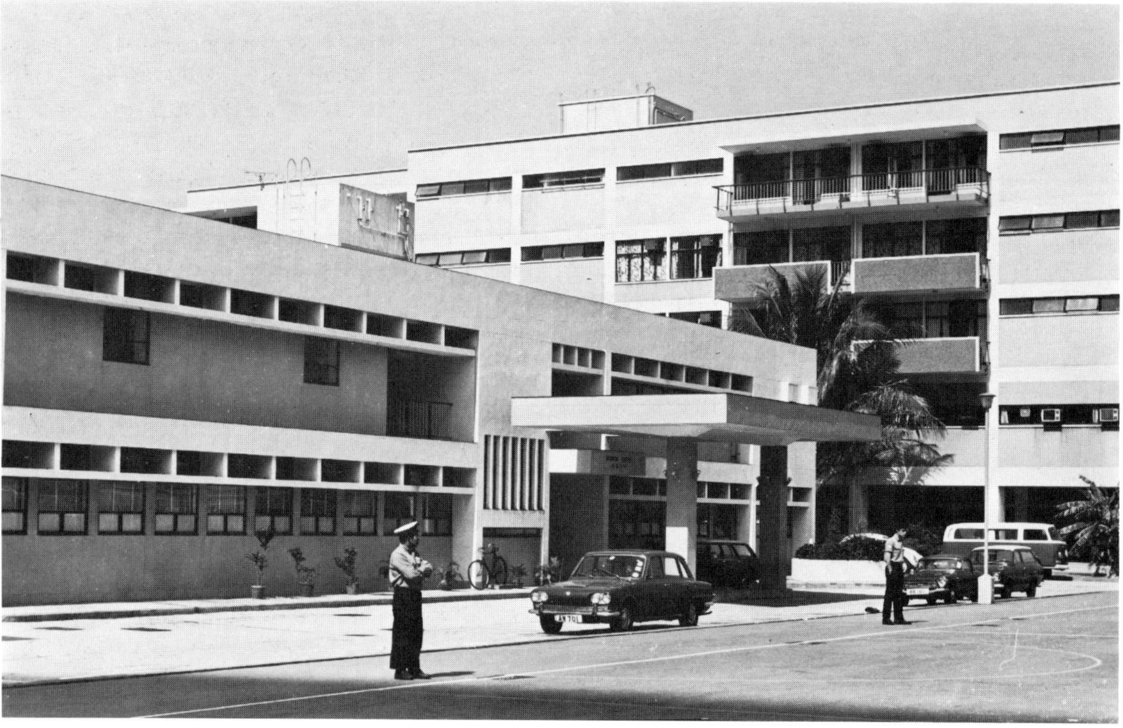

The Medical Centre in Tamar today—

Commodore involved in the negotiations.

It is interesting to note that quite early on efforts were made to obtain the services of nursing sisters for the naval hospital. At first it was hoped that these could be obtained from the civilian network but a letter of April 1895 from the Government Civil Hospital regrets that it has none to spare. The idea persisted, however. Writing to Vice-Admiral Sir Alexander Buller the Deputy Inspector states:

"The benefits of nursing sisters have been so amply proved and the comfort and help they give to patients is so generally recognised that the Royal Naval Hospital here must be considered "behind the times" without them. All the civilian hospitals in Hong Kong are well provided with them, and they stand the climate as well as other people."

He goes on to point out that building then in progress at the hospital could provide ample accommodation for two nursing sisters and the necessary domestic staff (a cook-boy and an amah at twelve dollars a month!). It was not until 1904 that sisters were obtained from the Queen Alexandra Royal Naval Nursing Service, members of which served continuously in the naval hospital until its closure in 1959. Three were working in the hospital when the Japanese attacked in 1941. Although it was hit by bombs, between sixty and one hundred times, the three officers survived. At the surrender they were transferred to Stanley Internment Camp, from which they were released in August 1945, to take passage home on H.M. Hospital Ship, Oxfordshire.

When it came to reorganisation at the end of the war, it was found that the fabric had suffered considerably: bombing, looting, vandalism and neglect had left only a shell which, it was felt, would have been more costly to resurrect than to start afresh elsewhere. Immediate Service needs were met by taking over two floors of the Queen Mary Hospital, while discussions were in progress as to the positioning of the new hospital. An excellent site at Black Links had been offered to the Navy in the 1920's and there were great hopes that this would now be utilised. Admiralty approval, however, was not forthcoming; inevitably, a compromise had to be reached. In November 1946 it was decided to take over the War Memorial Hospital on the Peak, which however, having been unoccupied for some time, required a fair amount of work. This was completed by 1948 and in January 1949 the Navy took possession of the new hospital on the transfer of all patients from the Queen Mary.

There was much of which to be proud: situated on the ridge leading to Mount Kellett, the site commanded magnificent views over the hills and valley to Aberdeen and the coast; the building itself, originally constructed as a nursing home, was spacious, capable of housing the full range of specialist departments among which the one hundred and fifty beds were divided, as well as providing accomodation for many of the staff. It is interesting to note that the quarters of the Sick Berth Staff were the two single storey buildings, only a few minutes walk away, previously used as the naval sanatorium. This particular one had been opened in 1889, though certainly with no great enthusiasm on the part of the Commander-in-Chief. However, it gained Admiralty approval and was certainly functioning in the mid 1920's; it seems to have been used as a refuge from the heat for personnel serving on the ships. A second sanatorium had been established at Mount Gough but seems to have had a shorter life, and was finally demolished in the typhoon of 1923.

The hospital functioned until 1959 when economy measures, decided on by the home government, forced its closure. Thenceforward, naval patients were sent to the British Military Hospital, formerly situated in Bowen Road and latterly in Wylie Road, Kowloon. Currently the health of servicemen and their families on Hong Kong island is looked after by naval doctors working in the Island Group Practice, situated in H.M.S. Tamar.

CHAPTER 11

## THE NAVY IN THE NEW TERRITORIES

In 1898 the Hong Kong government took over, on a ninety nine year lease, an agreed amount of land—what came to be known as the New Territories—on the Kowloon mainland. This was mainly for reasons of defence: China at that time was in a weakened state, a fact which had allowed Germany, France and Russia to obtain territorial concessions from her. Fearing that any further acquisitions by any of these countries would threaten British interests in the area, the home government determined to enlarge the size of the colony. Accordingly an agreement was reached with the Chinese for the lease of 355 square miles, extending from Deep Bay and Mirs Bay in the north to Lamma Island in the south. It was not to be supposed, however, that the transaction would pass off without incident. Naturally there were those Chinese who felt that their way of life might be threatened by the westerners, or, possibly, they were unwilling to exchange a remote authority for one much closer to hand. Misunderstandings by officials also played their part. At any rate, the Navy was called on to play a role—mainly a supportive one—in pacifying certain trouble spots.

The action began at midnight on April 3rd 1898 with a hurried note from the Governor, Sir Henry Blake (remembered today in "Blake Pier") to the Senior Naval Officer, Commodore Powell, asking for one of Her Majesty's ships to convey about one hundred soldiers to Pai Po Hu (the modern Taipo) where the Civil Secretary, six Sikh policemen and ten Chinese "braves" were holding out against a threatening mob. H.M.S. Whiting accordingly sailed at 3.30 a.m. having embarked the troops and the Commander British Forces himself, General Gascoigne. Arriving at Tolo harbour in Mirs Bay at 9.00 a.m. the troops were disembarked accompanied by a Lieutenant. In the rear followed H.M.S. Fame as dispatch vessel and two gunboats. It all turned out to be something of a non-event: a matshed had been burned down but the Civil Secretary and his party had been allowed to leave in safety. The general, seeing no reason for the troops to remain, signalled at 1.30 for the boats to be sent off to pick up the men; everyone was back home in time for tea! On the way to Mirs Bay H.M.S. Whiting had suffered some damage through striking a shoal, but General Gascoigne, in a letter to the Commodore, sportingly pointed out that all speed had been urged and that the mishap seemed unavoidable. The Commodore's comments go unrecorded!

Trouble of a more serious nature did, however, develop a few days later in the same area. About a thousand uniformed Chinese gathered on the hills round Taipo and began to fire down into the camp of the recently reinforced Hong Kong Regiment. H.M.S. Fame, ordered up in support, opening fire with its eleven-pounder guns, soon silenced the Chinese battery. Lieutenant Keys landed with a party of men and helped the troops to clear the hills. The incident was sufficient to make the governor decide that the Union Jack should be raised a day earlier than planned because, when the territory had passed into British possession, he could then act freely against any Chinese malcontents. And so, on the 16th April, the General, Commodore and Mr Lockhart (the Colonial Secretary) proceeded in H.M.S. Brisk to Taipo, where the ceremony was performed with no opposition. Since the Governor wished the flag to be shown also in Kowloon City, H.M.S. Peacock was towed there (not having had time to raise steam) to fire a salute.

May saw the final incident in the "campaign", this time, on the border. General Gascoigne asked Commodore Powell for naval help in the occupation of Sum Chun. This took the form of a naval brigade, commanded by Captain Clarke of H.M.S.

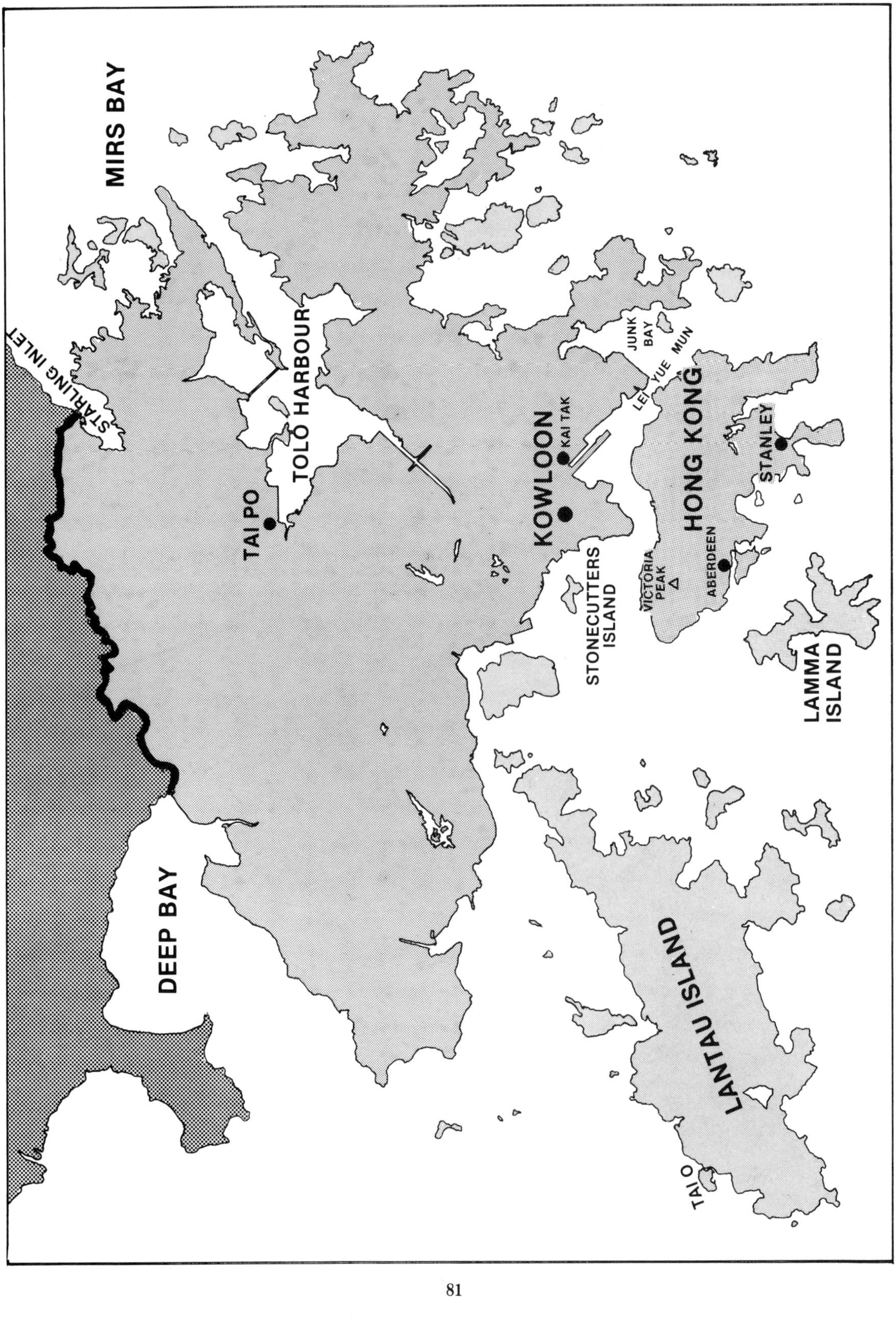
MIRS BAY
STARLING INLET
TOLO HARBOUR
TAI PO
DEEP BAY
KOWLOON
KAI TAK
JUNK BAY
LEI YUE MUN
HONG KONG
STANLEY
ABERDEEN
VICTORIA PEAK
STONECUTTERS ISLAND
LAMMA ISLAND
LANTAU ISLAND
TAI O

Undaunted. It consisted of one hundred bluejackets and thirty marines from his own ship, the same from H.M.S. Aurora and thirty marines from Tamar. They were taken up in the Swift and the Firebrand to Starling Inlet, where they disembarked and marched inland to meet converging columns from Taipo and Deep Bay. The Union Jack was hoisted at Sum Chun and a twenty-one gun salute fired. There was no sign of any Chinese. Everyone being now satisfied that the right sort of demonstration had been made, the troops and ships departed the following morning.

This series of events can seem unimportant because of the lack of action and the sense of anti-climax which that generates. Yet it is surely a case of "what might have been". The Chinese dissidents were real enough; the fact that they did not make their presence felt was due to the prompt action of the civil and service authorities, and to the efficiency with which their orders were carried out. The time when power changes hands is always a more crucial one; the process of consolidation is often much easier. The New Territories is important in present day Hong Kong because many industries and shipping firms are based there, and it provides valuable land in a colony always hard-pressed for space. What will happen when the lease expires in 1997 is not a problem which will trouble those who secured it at the turn of the century. Of course, the Navy would not claim that theirs had been the dominant role at that time. Indeed, Commodore Powell is on record as saying "The question of pacifying the newly acquired territory is a military one and should be left to the troops as much as possible", but from England came this tribute from the Secretary of State for the Colonies:

"I have pleasure in bringing to the notice of the Lords Commissioners of the Admiralty the prompt and efficient services rendered by naval officers trusted with the transport of troops."

A February 1897 photograph rather charmingly entitled, "Hosts and Guests". Ships of three navies are assembled in the harbour with the New Territories in the distance to the North. The dockyard buildings are in the immediate foreground.

CHAPTER 12

# GENERAL CONTRIBUTION TO COLONIAL LIFE

It is perhaps inevitable that the Navy should have had more impact on colonial society in the early days than it has had later. This is because the ratio of servicemen to civilians was much higher then than it is now, and consequently any personalities or "characters" among the former were more easily noticed in a European population which was both inward-loking and home orientated. Any society which numbered billiard tables among its first imports and which furnished its houses on the Peak in the latest London fashions, would obviously defer to Her Majesty's representatives abroad. Proof of the impact made by individual officers can be seen in the many place names with naval connections: apart from those already mentioned there is Mount Parker, named after Vice-Admiral Sir William Parker, Kellett's Island and Mount Kellett commemorating another admiral of the 1870's and Cochrane Street recalling the very popular Commander-in-Chief of the 40's, Rear Admiral Sir Thomas Cochrane who, it is recorded, entertained a great deal and was socially one of the best known personages of his time. The famous Hong Kong Club, founded in 1846, allowed naval officers membership at a fee of only three dollars a month. Indeed, it still extends its hospitality to members of the local Squadron.

Another feature of the early days was the public addresses presented to certain personages before their departure for home. The Senior Naval Officer, Captain McQuhae, was the recipient of one such in 1848. Sir Michael Seymour in 1859 went one better when he received not only a magniloquent address but two thousand guineas "for services rendered and to show respect." Nothing at all was done for the governor, Sir John Bowring, when he left a few weeks later! The best documented of these addresss is, however, the first one, written to Captain Charles Talbot of H.M.S. Vestal in 1847. The letter reads in part:

Dear Sir,

"Having just learned with regret that you are about to take your immediate departure from China (note, not from "Hong Kong") we hasten to express to you . . . how much we appreciate the kind consideration, the ready courtesy and the unremitting attention which you have ever displayed in your official capacity towards the mercantile community in China."

This extract gives us a clue to a further contribution the Navy made to the domestic life of the colony at its beginning, apart from mere socialising. The merchants would have appreciated the naval role because it was their shipping that was protected by the guns of the men-of-war from pirates and the junks of the Chinese emperor. Admiral Seymour was rewarded because he had vigorously prosecuted the "Arrow War" (started when a Hong Kong registered ship with a British captain, was attacked while alongside at Canton). Apart, however, from doing any fighting that was necessary, the Navy made itself useful to the mercantile community in other ways. For example, all the early surveys of the harbour and surrounding waters were performed by Royal Naval ships, among them H.M.S. Sulphur and H.M.S. Nassau. Until that was done, the China Sea was largely uncharted and was therefore dangerous for merchant shipping. In 1866 the first recorded levelling in the colony was performed by men from H.M. Surveyng Vessel Rifleman. A copper bolt was driven into a store house wall in the dockyard at a height of 17.833 feet above principle datum. Since then, all levelling in the colony has been based on this principle datum, which is approximately 3.9 feet below Mean Sea Level. There is a plaque in the naval base today which commemorates this.

Another service performed for the community until it was properly organised was the carrying of mail between Hong Kong and

**The arrival of the Duke of Edinburgh in HMS Galatea (centre) in November 1869. Just to the right is HMS Princess Charlotte. The hospital ship in the left distance is probably HMS Melville, sold in 1873 to pay for the new Royal Naval Hospital.**

**The Marine Police Station and the Signal Station with the Time-ball at Tsim Sha Tsui in about 1908. Attempts are now being made to save these buildings from demolition. The Time-ball idea had been earlier copied on board HMS Victor Emmanuel.**

the treaty ports in China, where the merchants were permitted to carry on their trade. There is a record of a letter written by the Postmaster in August 1847 to the Admiral's secretary, saying that he had heard that H.M.S. Vulture was leaving for Amoy, and could he have the time of its departure, please, so that he could make up the mail. These somewhat casual methods, however, could lead to problems: two years later the Postmaster was writing to the captain of H.M.S. Inflexible asking why some letters which it brought back from Shanghai were delivered direct "to certain mercantile firms" instead of going to the Post Office, this having been "to the serious detriment of the local Revenue!"

The ceremony and pageantry which are a part of the naval routine must have added a good deal of colour to life, again, more in the last century than this, when counter-attractions have developed so much. Whether it was swinging down the Praya on the way to exercise at Happy Valley, snapping to attention at a parade like that held in 1891 for the colony's 50th birthday, forming a guard of honour for a wreath laying ceremony at the Cenotaph, putting on a show of illuminations (as Tamar did for the coronation of Edward VII in 1902) or a searchlight display (as for George V's coronation), the Navy was involved in much of the formal life of the island in some way or other.

The Navy too has always been ready to render assistance in any domestic crisis, often brought about by natural disasters: a few incidents from recent times can be selected. Helicopters from H.M. Ships Eagle and

**The unveiling of the Cenotaph in 1923. The Hong Kong Club is to the right. The Royal Navy has always had a role to play in the formal life of the colony.**

Hampshire were used to reconnoitre and supply inaccessible areas cut off by a freak rainstorm of June 12th 1966; naval vessels assisted with the rescue of survivors from merchant ships during and after Typhoon Rose on August 17th 1971; Naval divers assisted government forces after a disaster near Kai Tak in 1965, and again combined with them in 1976 in the recovery of several million dollars worth of drugs. The Hong Kong Gazette of 1970 states that in the past year the Navy "contributed men and materials to installing or improving recreational and domestic facilities for the inhabitants of isolated villages and islands," a theme which recurs in 1976 when "in a neighbourly spirit the Royal Navy provided much help to villagers in the rural areas of the Sai Kung and Tolo peninsulas. Recent projects included the installation of generators and the provision of electricity for schools, villages and street lighting, the construction of a children's playground and the provision of dental treatment." Close links are also maintained with the Hong Kong Sea Cadets Corp and the Stanley Sea School whose boys are regularly taken to sea for training sessions.

**Royal Naval officers with village elders at the switching on of electricity to Leung Shuen Wan, one of the many community projects sponsored by the Navy.**

CHAPTER 13

## PIRATES ON THE CHINESE MAIN

Hong Kong island was taken over and colonised as a substitute for Canton. From the very first, therefore, trade was its mainstay and "raison d'etre", and any threat to the carrying on of that trade had to be resisted with the utmost vigour. The challenge came, not from the Imperial Chinese Government, as might have been expected, but from its less lawful members. Pirates had long been established on the China Seas. How best to deal with them was certainly a tricky problem for the Navy. Legally, it could certainly attack at sea, but not within coastal waters to a range of three miles. Thus, the geography of the locality, with its many bays, inlets and estuaries was heavily weighted in the pirates' favour. It was official policy that pirates could only be arrested if caught red-handed: when Captain Collinson of H.M.S. Plover (which has given its name to one of the most beautiful coves in the New Territories) commandeered what he took to be a pirate junk he was criticised for behaving aggressively. The civil authorities, of course, were always concerned with maintaining the best possible relations with the Emperor. Thus it was that chasing individual pirate vessels remained very much a hit and miss affair.

Paradoxically, as the pirates' organisation improved, so did the Navy's task become easier: in the 1850's and 60's set piece battles became more common, where naval skill and discipline told against sheer audacity. One such running battle took place between September 28th and October 3rd 1849 between H.M. Ships Columbine, Fury and Medea and the junks of the number one pirate chief, Chui Apoo, (undoubtedly the Henry Morgan of the China Seas), which resulted in the destruction of Chui's entire fleet consisting of twenty three junks, each armed with twelve to eighteen guns manned by 180 desperadoes. Two mini-dockyards on remote islands where the fleet had repaired and replenished were also destroyed.

A few weeks later, between October 19th and 22nd, the same officer, Commander J.C. Dalrymple Hay, led another attack. His assault force this time consisted of H.M. Ships Phlegeton, Fury, Columbine and a large number of officers and men from H.M.S. Hastings, the objective being the fleet of the number two pirate chief, Shap-ng-tsai. This resulted in the greater part of the pirate force being eliminated, enough anyway to make the merchants in their gratitude donate a "suite of plate worth £200 to each captain present at this action," which was a nice gesture although the money would represent only a small proportion of the value of cargoes saved.

Although this, as an isolated action, was undoubtedly successful, the naval authorities were not satisfied with the way the war as a whole was being prosecuted. Rear-Admiral Austin, for instance, in August 1851 made the accusation that petty piracy near Hong Kong was encouraged and fed through collusion with Chinese living on the island. The Governor, Sir George Bonham, had to agree with him but the civil government does not appear to have done anything at this time to check the liaison.

A third pirate fleet was destroyed by H.M.S. Rattler on May 10th 1853 and a fourth by H.M.S. Barracouta in 1854. Of the last named expedition we are fortunate enough to have an eye-witness account which gives an excellent picture of what it entailed to run a pirate fleet to ground.

### THE BATTLE OF KUHLAN

"On the 11th November (1854) the Barracouta, with two Imperial (Chinese) junks left Hongkong—(note the old form of spelling; the modern form was not authorised by the government until 1926)—to join the expedition sent against the stronghold of Kuhlan in the island of Tylo. We met H.M.S. Spartan and got an armed boat from her, and

**Landing seamen at the Pie-Ho River in 1859. Similar actions to this were fought along the China coast in the 1840s and 50s.**

anchored at night near our destination. Next morning at daylight we weighed anchor and soon joined the expedition, consisting of H.M. Ships Encounter and Styx, the U.S. steamer, Queen, and the Peninsula and Oriental Company's steamers, Canton and Forbes, with the Winchester's launch in tow, a very important force. The small steamers of light draught were necessary owing to the shallowness of the bays and the imperfect surveys of the island and soundings about Tylo.

On passing the island of Cowcock we descried three junks making all sail towards land, endeavouring to escape, their guilty conscience hurrying them. Faster—faster! Overboad went the guns one after another: anything to lighten them. Three boats were sent in chase from the Spartan and Barracouta with the Winchester's launch. Before the boats reached them the pirates had taken to the water and were now scampering up the hills, occasionally taking a rest and a last look at their junks and booty, all speedily committed to the flames.

We found that the large steamers could not get within five miles of Kuhlan, so it was determined by Captain O'Callaghan of the Encounter to proceed in the morning with the small steamers towing the armed boats of the expedition. Mr Sarratt of the Spartan spent the greater part of the night in a small boat sounding the bay of Kuhlan within musket shot of the piratical fleet.

Long before daylight on the morning of the 12th I was roused from my slumbers by the rattling of muskets and cutlasses, each man procuring his own arms. Our paddleboats with a 24-pounder howitzer in each, and pinnace with a 12 pounder, were ready alongside. Having dressed and swallowed a cup of coffee I took up my position in the pinnace, with my side-arms and surgical instruments. The small steamers towed the boats within 2,000 yards of the pirate fort and cast them off. In the meantime, the Portuguese war lorcha (a vessel with a European type hull but a Chinese rig), Amazona, armed with ten howitzers, sailed in and anchored within six hundred yards of the fort. She had just joined us from Macau.

The island of Tylo is about fifty miles to the south west of Hong Kong and twenty south of the Portuguese settlement of Macau. It is irregular in shape and very hilly, one range of hills extending the entire length on the north, another on the south, and between them a pretty valley winds round jutting rocks, spreading occasionally into broad paddy fields, and narrowing again for the passage merely of a fresh-water stream. At either end of the valley is a sandy beach, dry for a considerable distance at low water, and at full tide a large boat could get within two hundred yards of the shore. On the west, both ranges of hills jutted out for about eight hundred yards into bold headlands and formed the boundaries of a crescented bay; on the side of this an artificial basin with locks was formed for the reception of junks requiring repair, and near it, fifty piratical junks were at anchor.

An embarkment was thrown up from side to side, behind which twenty guns peeped from their embrasures; this protected the approach to the village of Kuhlan, which was built on a gentle inclination of the north side of the valley. The homes were very good and uniform in size and, for a piratical village, very neat. The pirate flag, black, waves from every eminence, and at each embrasure of the fort a man stood waving the flag of defiance.

Ere the armed boats had reached within range of the fort, the pirates opened fire with all their guns on the little lorcha. The latter treated them with contempt, not deigning to return a single shot till she received the signal from Captain O'Callaghan to "engage", when, with quickness and precision, she returned the fire. The line of boats also opened fire, advancing as they fired.

In the meantime, a party of marines under Lieutenant Burton landed on the north side, accompanied by a rocket party. Finding the fort getting too hot to be comfortable, the pirates gradually slackened their fire and a large number of them issued forth to meet the marines, who were steadily advancing on the fort. The boats, running aground, were left in charge of boat-keepers, whilst the crews, with musket and cutless, took to the water and were soon on shore.

About two hundred of the pirates retreated to a battery near a temple placed on an elevated terrace, whilst the rest scampered away in all directions, chased by bluejackets. Those in the battery, urged to fight by the exhortations of their chief who was arrayed in a robe of scarlet, maintained their position till fairly driven out by the bayonets of the marines.

The Portuguese and the Americans were conspicuous in the affray: the former attaching themselves to the marines who were about to scour the island. The bluejackets had taken possession of the junks and were preparing to set fire to them; a work for which they are pre-eminently adapted: they like to see a good blaze, it accords with their joyous dispositions. From behind rocks and shrubs gingall balls came dropping around in all directions.

The day grew hot and bright and as the marines with our allies passed to the other end of the valley, the firing became desultory. I accompanied this party. We examined each house and hamlet we passed; some of the cottages were filled with fowl and pigs, scarcely a soul was to be seen, save an old man or woman too old to run. After some heavy marching we reached a creek communicating with the sea. In it were moored twelve boats resembling whale boats, but larger, and capable of living in any sea; in such as these the pirates boarded the "Caldira". (This is a reference to a recent piracy.)

A handsome temple with a few straggling cottages stood on the beach close to the creek. The cottages were filled with nets, ropes, hawsers and many articles plundered from honest traders. The interior of the temple was richly decorated with banners of fine silk, gongs, tom-toms, spears and a fair sprinkling of josses on an altar gaudily painted and gilded. Anxious to propitiate Buddha they had a fair repast for this divinity of oranges, cake and tea; and incense was burning before him. On this occasion all their gods appeared to have deserted them, vanishing in flames, which in a short space of time left but the bare walls standing.

There was a halt to rest and refresh the inner man with whatever our haversacks produced. A portly editor, a little out of breath, with sun-burnt visage, sat apart on a bank with a large joss, his trophy suspended from his neck, busily engaged in committing to paper the events of the day; with laudible zeal he always tried to witness any little affair of a warlike nature, and even the bright rays of the sun failed to quench his literary ardour.

A most amusing incident occurred during the day. The officers, on leaving their respective vessels, provided themselves with haversacks and whatever food the ships afforded: some were satisfied with a little biscuit and cold pork, with a flask of rum, others with something more delicate. A Mr B. had provided himself with a roast fowl, cold ham, case of cheroots and a flask of brandy; but in his anxiety to land from one of the boats he took up the wrong haversack, containing a humble supply of biscuit and pork. A group of officers were collected together lying at ease under the shade of some wild pine, when Mr T., to our infinite amusement, drew forth from Mr B.'s bag the fowl and concomitants. The worthy owner was far away on some other part of the island, so we drank his health in the brandy and very much relished the cheroots—and the joke.

We resumed our search and returned to the place of embarkation about four o'clock where we found the force with a fair display of prisoners. Presently we were joined by Captain Rooney of the Caldira armed to the teeth, laden with some of his discovered property, and driving before him four pirates. We enquired how he managed to capture so many. "Faith," replied the warm-hearted Irishman, "didn't I surround them?"

Captain O'Callaghan publicly thanked Lieutenant Scarnachia and the gallant crew of the Portuguese lorcha Amazona for their conduct during the day. We then embarked on our boats, having to wade out to them, and in an hour were comfortably located on board the ships, and steaming away from the much humbled stronghold of Kuhlan. It is needless to state that for a considerable period afterwards not the sound of a pirate's voice was heard on the water; those who escaped from Kuhlan joined the Chinese rebels.

One fine young American marine, belonging to the U.S. frigate Macedonian lost his life, some say by accident, others from a gingall ball. We conveyed his remains to Hongkong in the Barracouta. On the 14th (the same day) we left Hongkong and cruised amongst the Ladrone Islands for some days, but saw no more of our piratical friends."

And so ends this fascinating account of the first expedition against Kuhlan. That it failed to subdue the pirates completely is shown by the need for another expedition the following year.

With the arrival on the station of Admiral Sir James Stirling the war against the pirates was carried on with increased vigour. Having asked for shallow-draught steamers which could pursue their quarry inshore, eight small gunboats were sent out from home in 1856. Most of the new arrivals went into action between September 1858 and March 1859, not only capturing upwards of thirteen junks but destroying a pirate village, battery and slipways. The ease with which the pirates replaced lost vessels should not be attributed solely to collusion; threats and intimidation were also used. Admiral Stirling also began to organise convoys, whereby a number of merchant ships were escorted together to their destination under the protective guns of the sister service. Further operations were necessary in 1857 and '58 which checked, but did not end, the evil.

During the governorship of Sir Hercules Robinson (1859-65) the character of piracy changed in the sense that no longer did marauding fleets roam the coastal waters, looking for stray merchantmen, but attacks which seemed to have been carefully planned were carried out on selected targets. The Governor wanted the Commander-in-Chief, Admiral Kuper, to make four gunboats constantly available for anti-piracy work. The Admiral, however, took a much different tack: he felt that the civil government itself should do much more to control shipping in the harbour. With no proper system of licensing, innocent looking merchant ships could be, and were, turned into pirate vessels in one of the many quiet coves of the mainland. There was the problem too of the Chinese living on the island, since information received suggested that pirates were easily able to buy what arms and ammunition they wanted locally. In sum, the admiral felt that his ships were being called on to do a job which an efficient colonial police force should be performing. In government circles the truth of what he said must have been acknowledged since the new governor, Sir Richard Macdonnell (1866-72) began carrying through certain reforms: all ships now had to be registered, their movements within the harbour and their anchorages were controlled by the government, while all junks entering or leaving were inspected before being given clearance.

These reforms, then, mark a change in the official attitude which had been that pirates were solely a naval problem. They depended for their efficiency on a strong marine police, which was what the Navy had been asking for for some time. The efforts Governor Macdonnell made to bring this about lie outside the scope of this book, but certainly, in his time, piracy in local waters ceased to be the chronic problem it had been. It did not die out completely, of course. Isolated incidents continued into the present century: as late as 1925 an officer is writing in his diary that "pirates in the Delta (of the Pearl River) have been fairly active lately and they have attacked one or two Chinese ships bringing food to the colony." Naval vessels still went on "pirate patrol" but increasingly the problem was left to the Marine Police, and with the growing sophistication of that arm of the colonial government, it was perhaps a fair solution.

## SHIPS ASSOCIATED WITH HONG KONG

All the ships which were on the China Station in the very early days were wooden sailing ships, usually 3rd rates, carrying between 60 and 74 guns. An exception was the 1st rate, H.M.S. Melville, which was at one time flagship of the East Indies Squadron but ended her days as a hospital ship, being sold in 1873 to pay for the new Royal Naval Hospital. A permanent China Squadron was established in 1844, but for many years this consisted of scarcely more than six ships. This number was, after all, enough to tackle the pirates or to enforce the trading rights so dearly bought at the Treaty of Nanking. The treaty ports—Canton, Amoy, Foochow, Ningpo and Shanghai—were normally, in the 1840's, given a brig each, except for Canton, the nearest and busiest, which usually had a steamer. The brig was a small, light vessel, ideally suited to coastal work. A frigate and a steamer were stationed at or near Hong Kong itself. Another type of ship seen in these waters was a corvette, really a French word, meaning a vessel with one tier of guns.

The real workhorse of the Navy in Hong Kong in the nineteenth century was, however, undoubtedly the gunboat, which meant rather more than simply a boat with a gun on it. The term was applied to various types of steamer, all well armed and suitable for river work. Not long after the Crimean War (1854-56) larger gunboats were built & specified as "gun vessels."

Apart from the Audacious and Victor Emmanue8, all the ships mentioned as stationed in Hong Kong in the chapter on the year 1878 were of this class, although in newspaper accounts they are frequently referred to as gunboats.

Of the larger ships a few are worthy of special mention as having close connections with the station over a number of years.

### H.M.S. PRINCESS CHARLOTTE

Launched at Portsmouth on September 14th 1825, the Princess Charlotte was a first rate vessel of 2,443 tons and 104 guns. The Princess Charlotte after whom she was named was the daughter of King George IV (better known as the Prince Regent) and Caroline of Brunswick. Theirs was a disastrous marriage, the couple separating a year after the birth of their daughter in 1796. Charlotte grew up a difficult child but found happiness in her own marriage to Leopold of Saxe-Coburg. She died in childbirth, however, in 1817. Her grieving father erected the very fine monument to her which can be seen in St George's Chapel, Windsor.

**H.M.S. Princess Charlotte. It was an Admiralty regulation that hammocks should be aired when in port.**

HMS Victor Emmanuel 1855.

Until 1837 the Navy Lists show the ship stationed at Portsmouth, not yet in commission. Although such a situation may seem strange to the modern mind, it was not so at the time. In the post-Napoleonic era the dockyards were continuing to turn out large numbers of wooden sailing ships, but there was simply not enough work to go round, and perhaps not enough money either as, in peacetime, the government always cut back on military spending. So there she lay until 1837, when she was worked up to become the flagship of Admiral Sir Robert Stopford, Commander-in-Chief, Mediterranean. It was then that she gained her single battle honour, at the bombardment of St Jean d'Acre on the 3rd November 1840. After service in the East Indies she was sent to Hong Kong in 1858.

### H.M.S. VICTOR EMMANUEL

Laid down in the Pembroke Dock in May 1853, her original name was Repulse but this was changed in 1855 while she was still on the stocks. Almost certainly, this was intended as a compliment to King Victor Emmanuel II of Sardinia who, in 1861, became first king of a united Italy. Classed as a screw ship of the second rate, she was 230 feet in length and 55 feet broad, carrying between eighty and ninety guns. Like Tamar, Victor Emmanuel was used as a troop ship to Africa and the Far East before arriving in Hong Kong on November 4th 1874. There seems to have been a hand-over period of a month when both she and the Princess Charlotte are listed as receiving ships, and then on December 11th the Victor Emmanuel appears in the shipping lists as the sole receiving ship, in the charge of Commodore J.E. Parish (after whom Mount Parish is named), her armament reduced to a mere two guns. Over a period of twenty-one years the ship must have become a well-known feature in the harbour: from her masthead flew the commodore's broad pennant, from her decks salutes were fired in honour of arriving and departing warships, from its offices was conducted the day-to-day business of the Navy in the Far East. As the photographs show, her upper decks were covered with roofing, for extra space and to give protection in bad weather—and which certainly leant her a distinctive silhouette! In August 1897 H.M.S. Tamar arrived but did not immediately take over her new role as she is listed as "Receiving Ship in Reserve" until the end of the year.

**H.M.S. TAMAR**

The ship from which the present naval establishment derives its name was built at Millwall and launched at the beginning of June 1863. Of about 3,650 tons, she began her maiden voyage on January 12th 1864 as a troop-ship to the Cape and China. At this time, the Royal Navy had not yet fully changed over to steam power, consequently Tamar was originally "barque-rigged", meaning that it had three masts, the fore and main masts being square rigged and the mizen, fore and aft rigged. Later, however, it was equipped with a steam engine, giving a speed of about twelve knots. The early photograph of her shows her with funnels, but she was later re-boilered and the funnels reduced to one.

Tamar first visited Hong Kong in 1878 with reliefs for the crews of the gunboats. Before that she had seen service off the Gold Coast, taking troops for the Ashanti War. It is reported that one death and one birth occurred on the passage out from Portsmouth which took seven weeks. Her stay at this time was short for she soon departed for Singapore to pick up the 74th Highlanders who were to relieve the 28th Regiment then stationed in the colony. She was to return only once, in 1886, before her final return in 1897, when she relieved Victor Emmanuel as receiving ship. At first, the ship was moored at a buoy in the harbour, but from 1913 she was kept alongside the west wall of the dockyard's basin, with only occasional sorties for docking or to test her general sea-worthiness.

The First World War scarcely affected Hong Kong; it was very different with the second. When it became clear that the Japanese attack could not be halted, orders were issued that any type of vessel which could be useful to the enemy was to be put out of action and sunk. Tamar was one of the first to go. As Commander Crowther states in his account, she was taken out of the basin on December 8th and sunk at the buoy on the 12th. The actual scuttling was not as easy as might appear owing to the superstructure (which had been built over her decks to provide administrative accommodation) becoming airlocked. For some time a sizable portion remained resolutely above water, until the Royal Artillery was called in to finish off the job, a sad end for a ship whose name, however, is sure of a lasting remembrance as it has been given to the naval establishment.

The ship's crest, still in use today, is based on the coat of arms of the county of Cornwall, since the River Tamar divides that county from Devon. To have taken the whole arms would, of course, have been a breach of copyright and accordingly the base of the crest of H.M.S. Tamar reproduces only ten bezants (roundels) instead of the fifteen found in the arms of Cornwall. The wavy lines are self-explanatory.

The figurehead, carved in wood and dressed in flowing robes possibly represents Tamar the sister of Absalom mentioned in Samuel Chapter 2 in the Bible.

### OTHER SHIPS

The three receiving ships were those most closely connected with Hong Kong but, over the years, large numbers of naval vessels have called at the colony, or been stationed there, for long or short periods. A short survey of the type of ship seen in these waters and of the numbers in which they came might therefore be appropriate.

Looking at the nineteenth century Navy Lists, a certain pattern does emerge: at the base there was always the receiving ship and a despatch vessel, fairly often too the flagship (more usually to be found to the north) and a varying number of gunboats, about ten normally, of which four or so would be at Hong Kong while the others cruised the China Seas, visited the various treaty ports, such as Shanghai or Foochow, to protect British trading rights there, to pursue action against pirates and generally to show the flag. It was not until the last decade of the nineteenth century that the Royal Navy arrived in greater numbers in response to the political situation in the east. China's weakness and internal divisions (the major Taiping rebellion, lasting from 1850 to 1864, caused twenty million deaths) made her very liable to attack from Japan, Russia or France. This closing-in of the major powers is known as the "Battle of the Concessions" and it is therefore not surprising to see the Navy reinforced to protect and promote British interests. There were, then, on the China Station in 1900 three battleships, fifteen cruisers, six sloops, ten gunboats, five shallow draught steamers for river service, six torpedo boat destroyers (used as tenders), a store ship and a despatch vessel. Five years later and there is little change, except that a definite Cruiser Squadron of seven vessels has been created, and the number of T.B. destroyers has doubled; of about 350 tons, these had a lieutenant in command (as did the gunboats), with an engineer lieutenant, sub-lieutenant and gunner in addition. And so the complement remained up to and including the First World War, when a force of eight destroyers made its first appearance in these waters along with three submarines.

Naval routine in the 1920's and 30's seems to have centred on spending the winter months in and around Hong Kong and the summer in north China, often at the island of Liukungtao or Wei-hai-Wei as it was known, about ten miles from the Chinese mainland and with excellent anchorage facilities for the Fleet. Ashore, there was a cinema, football grounds, a canteen, formerly the Dowager's Empress's Summer Palace, several married quarters and not very much else. Amusements tended to be home-made and none the worse for that: the China Squadron Athletic Sports were held there in July, the Pulling Regatta was a feature in August while there was even a Golf Competition during September. It is difficult to say whether these were annual events but diaries of the time certainly contain many references to many happy hours of recreation, whether at Wei-hai-Wei itself in the more formal sports or at one of the surrounding islands like Sun Island, used for banyans and picnics.

A well known character at Wei was Sin Jelly Belly, who seems to have been "Mr Fix-it" for the island, overseer of all the traders who had passes to go on board in the dog watches. During the winter months, when the sea round the island is frozen over, the people there used to spend their time making model boats, baskets and such like in preparation for the return of the Fleet in the summer. Although naval ships did visit the island during the winter, this seems to have been less common.

**HMS Magnolia in the graving dock. The sturdy character of the gunboat is well illustrated in this photograph.**

The return trip to Hong Kong was usually made via several "good run ashore" ports either in Japan or down the east coast of China. One of the ship's company of H.M.S. Dorsetshire, on a visit to Yokohama in 1936, remembers the ship being a mass of flowers from bow to stern as she lay alongside and remembers too hundreds of boxes of matches being thrown onboard as entrepreneurs on the quay tried to advertise various shows, clubs or entertainments. Free rail passes were also often issued to Royal Navy personnel for the duration of the ship's stay. Another highlight which always features in memoirs was the voyage from Shanghai down the several hundred miles of the River Yangtse to Hankow.

In the period 1920 to 1926 there operated from Hong Kong the 5th Cruiser Squadron, usually five vessels, one of which flew the flag of the Vice-Admiral, Commander-in-Chief, the 4th Submarine Flotilla (two depot ships, a tender and twelve submarines) and the gunboats, which were divided into the Yangtse Flotilla of ten (one flying the flag of the Rear-Admiral Yangtse) and the West River Flotilla of five vessels. Very often a minesweeper and survey vessels were also present. 1927 brought the arrival of the 8th Destroyer Flotilla (flagship plus eight), which remained until 1939, but apart from that, the list of the early 20's is essentially what was based in Hong Kong till the outbreak of the Second World War. In the 30's an aircraft carrier was present, and in 1939 "sloops" appear to have been re-classified as "escort vessels", but these are minor changes in what amounted to a very considerable force indeed. It would not have been possible to support such a group had not a major dockyard been constructed at the turn of the century, as described in an earlier chapter.

Immediately after the end of the war, the Commander-in-Chief, British Pacific Fleet, had his headquarters at Hong Kong and his flag ashore, but in October 1948 this changed: the headquarters of the Commander-in-Chief, Far East Station (as it was now called) was moved to Singapore, with Hong Kong being designated as the "Operational and Training Base" (a title which was dropped in April 1954). This change reflected the increased importance of Singapore as a naval base (it was established in 1934). The flotilla based at Hong Kong, which was certainly in being in 1949 and continued certainly till after 1954, consisted of smaller craft, Harbour Defence Motor Launches, Motor Fishing Vessels, fitted with types of mine-sweeping gear, and landing craft for which the dockyard carried out all necessary engine changes, alterations and additions. The flotilla performed boundary patrols and stop and search operations on craft suspected of carrying contraband or illegal immigrants. Although units of the 5th Cruiser Squadron, the 8th Destroyer Flotilla, the 1st Escort Flotilla and the 3rd and 4th Frigate Flotillas (or Frigate Squadrons as they were known after 1952) continued to call in for "Docking and Defects" or leave and recreation, there is not the concentration of force that there had been in pre-war times.

The 3rd and 4th Frigate Flotillas for a period in the 1950's each consisted of four ships, each commanded by a captain. The former was made up of H.M. Ships Crane, Modeste, Amethyst and Opossum, Crane being the Flotilla leader, while the latter's ships were "Bay" Class Frigates, including St Bride's Bay and Cardigan Bay. The leaders of the flotillas were easily recognisable from having a large black band around the top of the funnel and from the absence of the pennant number on the ship's side. Of these ships H.M.S. Amethyst is perhaps the best known, the ship caught in the crossfire of the Nationalist and Communist Chinese forces in 1949. Her celebrated dash down the Yangtse has been well recorded as has been the presence of Simon, the only "V.C." cat who, apart from keeping down the mice and vermin on board, endured some heavy shelling.

It is difficult to give a typical programme for ships in these days but it does seem that the earlier pattern was followed, with the summer months being spent to the north, often in Japanese waters. Many visits were paid ashore and a climb up Mount Fuji seems to have been a "must" on many a sailor's visit to the land of the rising sun. Visits would have been made south too, to Singapore, and for the annual SEATO exercise involving many Allied navies. At Hong Kong three month refits would be carried out, followed by work-ups and trials in the Mirs Bay—Tolo Harbour region. During re-fits the Fleet Accommodation Base was used, dormitory type quarters for about sixty, situated on the left hand side of the main entrance to the Dockyard. H.M.S. Tamar was also used and the China Fleet Club if necessary.

At this time, as Hong Kong did not have a guard ship, the "Emergency Destroyer" routine was employed. All ships in the area were classified as "Hong Kong Defence Force" and came under the operational control of Commodore Hong Kong. To conserve fuel all warships would, if alongside the jetty, shut down the boilers. One ship, a frigate or destroyer, would be at immediate notice for sea and would normally have moved to a buoy in midstream, for speedy action in case of an emergency. After a twenty four hour spell of duty she would return alongside and her relief would go out to the buoy.

One incident involving HMS St Bride's Bay when she was Duty Destroyer is worth retelling. One of the regular ferries from Hong Kong to Macau had just cleared the harbour entrance when she was intercepted by Communist gunboats who signalled their intention to send a boarding party in search of an old Nationalist Chinese general travelling in the ferry whom they wanted to transfer to one of their own ships. The ferry having warned Hong Kong by radio, the St Bride's Bay steamed out at full speed, firing four inch shells directly ahead of the gunboats which then made off. From accounts of those who served in the East at this time such an incident was not unique.

H.M.S. Unicorn perhaps paid more visits to Hong Kong than any other British warship in the Far East during the Korean conflict. She was a huge Aircraft Repair Vessel, on the "tramlines" run—Singapore, Hong Kong, Iwasuni, Sasebo, Kure, then back to Singapore for another cargo of stores. On one particular voyage, transporting about ten Canberras to Iwacuni for the R.A.F., she received a distress call from a British merchant ship in the Formosa Straits which was apparently being molested by a Nationalist destroyer in the vicinity of Round Island. Once on the scene, however, the appearance of the Unicorn, looking very similar to an aircraft carrier about to launch its planes, quickly encouraged the Nationalist ship to beat a hasty retreat.

Both the Korean conflict and the Suez crisis upset normal naval routines as ships were diverted to respond to the changing political situation. Both the Unicorn and H.M.S. Crane spent much time up the west coast of Korea, giving gunfire support to the Army ashore.

One interesting ship which visited Hong Kong in the 1950's was the old aircraft carrier H.M.S. Venerable, which had been with the Fleet on its return to Hong Kong at the end of the war in August 1945. In the 50's her name was the "Karel Doorman" for she had been sold to the Netherlands by the British government. The Dutch navy eventually sold her to Argentina, and she is now serving as the only aircraft carrier in the Argentinian navy. Her name, which became well known during the Falklands War, is, of course, "The 25th of May".

In October 1958 the 120th Minesweeping Squadron arrived, to be under the control of the Commodore Hong Kong, consisting of H.M. Ships Damerham, Darsham, Davenham, Glentham and Hovingham. These were reinforced in October 1962 by the 8th Minesweeping Squadron of Dufton, Penston and Lanton. With Hong Kong as their base, these minesweepers carried out many Borneo patrols until the Confrontation in Malaysia ended in 1966. The following year, as part of the United Kingdom's defence economy measures, they were transferred to the Inshore Flotilla at Singapore. The plan was that Hong Kong would be supplied, when necessary, with ships from this flotilla on a rotational basis and, of course, every major warship in the Far East such as H.M. Ships Eagle, Albion, Intrepid, continued to call in, along with their support frigates and R.F.A.'s. A guardship, usually a frigate, was stationed in the colony, the last of these being H.M.S. Chichester, from 1974 to March 1976.

In 1969 there was another change of plan: as part of the redeployment of the Far East Fleet, the 6th Mine Countermeasures Squadron, compising H.M. Ships Maxton, Bossington, Kirkliston, Sheraton and Hubberston was permanently transferred to Hong Kong. The following year they adopted the Chinese dragon as their squadron badge, displayed on the funnel of each ship. Previously the squadron badge had been a black foot motif, deriving, by tradition, from the marks left by the stokers on deck during coaling, in the days of steam. 1971 saw the replacement of Bossington, Hubberston and Maxton by Beachampton and Yarnton, while in the next year, Wasperton, Wolverton and Monkton completed the new-look Hong Kong Squadron, re-named the 6th Patrol Craft Squadron, since the minesweeping equipment had been removed from each vessel. The arrival of the squadron entailed a build-up of the necessary technical personnel in the Tamar base. Initially, the ship's programme was centred on patrols round the colony's many islands, providing assistance, if necessary, to the Marine Police, maintaining weapon efficiency and carrying out "hearts and minds" operations, particularly among isolated communities. The force may seem small but, interestingly enough, it corresponds almost exactly to what was present for most of the nineteenth century: certainly,

HMS Glentham of the 120th Minesweeping Squadron is dwarfed by visiting US Naval Ships.

office blocks have replaced the receiving ship, the telephone and wireless have superceded the despatch vessel, but the patrol craft is really only the gunboat under another name. Over 140 years after the establishment of Hong Kong, it would seem that operational requirements have come full circle.

**The Chinese Dragon—funnel emblem of ships which have served in Hong Kong since 1970.**

**The frigate CHICHESTER — leader of the Hong Kong Squadron alongside HMS TAMAR. The old dock entrance, behind her bow, can easily be seen.**

CHAPTER 14

## RECENT EVENTS

The Royal Navy's task in Hong Kong, as has been suggested by the preceding chapters, has always been to aid the local government in maintaining the political equilibrium, to be prepared for the outbreak of a limited war in the eastern hemisphere and to promote good community relations. This triple commitment, however, came to be overshadowed in the late 1970's and early 80's by the pressing problem of illegal immigrants, or "II's" as they are known locally, who arrived from Vietnam and mainland China in increased numbers drawn by the bright lights and material prosperity of a colony where they literally believed the streets were paved with gold. In one way this was no new thing: refugees had flocked to Hong Kong in 1949 after the Communist victory in China and again in the 1960's during the Cultural Revolution, but what was vastly different in the late 70's was the sheer size of the numbers involved. The authorities were aware that those caught represented only the tip of the iceberg: in 1978 some 8,000 were caught but an estimated 28,000 got through, while in 1979, 90,000 were caught but an estimated 108,000 were not. Obviously, such figures represented a real threat to the stability of a society already under pressure to meet its housing, health and educational targets. Forces would have to be mobilised to meet this unusual invasion, this invasion of civilians, mostly young men, but including many women, children and babies, usually unarmed and often pathetically ill-equipped for their journey, but sometimes desperate and always determined.

The assault came by land—across the barbed-wire fence erected along the Chinese border from Sha Tau Kok in the east to the

**The Hong Kong Squadron during the 1970s and early 1980s HM Ships Wasperton, Beachampton, Woolverton, Monkton and Yarnton.**

A typical boat load of illegal immigrants—

marshes of Deep Bay in the west—and by sea, from quiet bays and inlets where a junk or rowing boat could shelter before "risking all" in the final dash to reach the promised land. Against this threat the Navy mobilised about seven hundred officers and men, half of them Locally Entered Personnel, deployed mainly in the task of patrolling the eastern and western approaches to Hong Kong. The essence of patrol work is patience in the hours of waiting, alertness at the slightest suspicious sign, and diligence in investigation. This work was done by the five Ton Class converted minesweepers and by the ex-dockyard tug, Clare, whose job it was to stop and search any suspicious-looking vessel, such as one not displaying the required lights. It is, however, difficult to distinguish between a junk returning home with its holds full of fish and one which might be carrying a less legal cargo, and so the task demanded a certain feel and flair and the exercise of intuition, built up over several months' experience. That this method can be very successful is shown by the fact that during one weekend in 1980 H.M.S. Wasperton alone picked up four hundred and fifty II's. Once a boat had been discovered, the patrol craft came alongside and a Cantonese-speaking member of the crew addressed the silent, wide-eyed occupants by loud-hailer. They then clambered onboard, where they were first searched, then made to sit on the deck in groups under guard. Having been given something to eat and drink, they were then handed over to the Marine Police to be repatriated to China. Much of the action took place at night as most immigrants prefer to make their bid under the cover of darkness which can create problems for the Navy, as many of the craft in which they arrive—some little more than rowing boats, with a sail, quite open to the elements—were so small as scarcely to show up on a radar screen. These frail vessels often proved inadequate and it was the sad task of the patrols to pick out from the sea the bodies of those for whom it had all ended in disaster.

A new dimension was added to this nautical game of hide and seek with the introduction of speed. Those immigrants willing and able to pay an exorbitant price (approximately £2,000 in 1980) were transported from Macau (the Portuguese colony on the Chinese mainland) across the South China Sea in high-powered speedboats which could easily outstrip the patrol craft in any race. The Navy's

**Not everyone made it. A boat load of 'illegals' is taken in tow. The body of a fellow 'hopeful' floats astern.**

**A wide range of "equipment" was used in the hunt for an ever increasing number of illegal immigrants.**

The fast Patrol Boat HMS SCIMITAR was brought from the UK to assist in the hunt—to chase those using fast pleasure craft to run into Hong Kong waters.

**The major proportion of "II hunting" was carried out by the patrol craft however . . . no vessel was safe . . . Sampans to luxury yachts were all possible ferries for illegal immigrants. Here the Patrolcraft HMS Beachampton stops a weekending yachtsman as he returns to Hong Kong—just to check!**

**An SRN6 Hovercraft arrives at its base at Stonecutters Island after a night patrol in local waters.**

answer lay in its two 50 knot SRN 6 hovercraft and in the 40 knot fast patrol craft H.M.S. Scimitar, the latter berthed at H.M.S. Tamar, the former at Stonecutter's Island which frequently resounded to the roar of their engines as they bounded across the harbour ontheir way to the "fishing" grounds. Working closely with the patrol craft on duty and with the vessels of the Marine Police, these could easily dash to the area of a sighting and pursue the quarry. They were frequently helped by a helicopter fitted with the xenon-arc 65 million candle power searchlight, Nite Sun, which, trained on the speedboat driver, dazzled and disorientated him, causing him either to lose his sense of direction and go round in circles, or to cut his engines and give up the game. Occasionally a driver would lose control completely and smash himself, his boat and its occupants to pieces on the foreshore. From a crash at such speed there were few survivors.

Also involved in this battle of wits with the "Macau Runners" were the Royal Marines of No. 3 Marine Raiding Squadron in their Rigid Raiders. With great physical resilience, for the Raiders provide little protection from wet and bumpy conditions, the Marines would lie in wait night after night in quiet bays, waiting for the chase to begin. When it did, their role was often to join with the launches of the Hong Kong Marine Police in forming a moving funnel shape down which the speedboat was driven, with the object of diverting it from its intended course into the harbour. Although by day the Marines were accustomed to stopping and boarding junks, this was impossible by night, the boats operating

**Rigid Raiders prepare for another night patrol from Stonecutters beach.**

at high speed. Consequently the aim was either to force the speedboat out to sea where it would eventually run out of fuel, or into the waiting net. It was this type of action which enabled the success rate of the speedboats—high in the early days—to be severely reduced as the campaign went on.

It was towards the end of 1980 that the situation altered radically: the flood became a stream then a trickle. A mere handful of immigrants attempted to enter the colony and that was every week, not every day. The new situation was partly due to a change of policy by the Hong Kong government which previously had operated the "touch base policy", by which once an "II" had reached urban Hong Kong he was given an identity card, which ensured his right to stay. This was done to prevent criminal elements operating blackmarket or blackmail rackets. When the government abandoned this policy it meant that everyone in Hong Kong had to have an identity card and employers engaging staff without one were subject to a heavy fine. An additional factor in causing the reduction was probably the attitude of China which may have been doing more to prevent the escape of its citizens. The 42nd Army, some 20,000 men, was deployed along the land frontier and pressure too may have been put on the leaders of communes to see that their members did not make any abrupt departures. Since the "Touch Base" policy ceased in 1980 total arrests of II's dropped from about 90,000 a year to 10,000. The projected total for 1983 was 6,200.

The lessening of the refugee problem meant that the reinforcements called in to deal with it could be sent home: Scimitar was returned to the United Kingdom in 1981 while the Hovercraft unit ceased to be operational in April 1982 and was disbanded in September of that year. Although the Royal Navy continues to support the Royal Hong Kong Marine Police in anti-illegal immigration patrols, to which end a patrol craft is continually at sea, there is now more time for visits—pairs of ships of the squadron have continued to deploy away from the territory twice a year, visiting ports between Japan and Singapore—for both exercises and training. The process of modernisation also continued with the replacement, in 1984/5, of the thirty year old Coniston-class patrol craft with the new Peacock-class. These new ships are 75% funded by the Government of Hong Kong for both their construction—and operation. They are now in service in local waters—their long term role remains to be seen, as Hong Kong heads for 1997 when the lease on the New Territories expires—and the colony is returned to mainland China. In the interim the Royal Navy remains willing and able to fulfil whatever role it may be given in Hong Kong. Whilst generations of Naval personnel look back on times past spent in the area, it appears time is short for the current generation of naval men to serve in this, one of the Navy's most popular drafts.

**A Royal Marines Rigid Raider and a helicopter equipped with "Nightsun" posed a potent threat to the fast "Macau runners." The Nightsun would dazzle the cox'n who would often drive his craft ashore in an attempt to avoid capture.**

**The latest "secret weapon" in the hunt . . . A 55 knot Fast Pursuit Craft—normally only seen at night in Hong Kong waters.**

**End of an Era . . . The Patrol craft HMS Wasperton lies alongside as the new Patrol craft arrives on station.**

## The Old

**Whilst Traditional local craft continue to ply the local waters. .**

## The New

**HMS Plover heads the "new look" squadron into harbour.**